The Sun-heated
Indoor/Outdoor Room
for People, for Plants

The Sun-heated Indoor/Outdoor Room for People, for Plants

by JACK KRAMER

Drawings by Adrian Martinez

CHARLES SCRIBNER'S SONS
NEW YORK

Copyight © 1975 Jack Kramer

Library of Congress Cataloging in Publication Data

Kramer, Jack, 1927-
  The sun-heated indoor/outdoor room
for people, for plants.

  Bibliography: p. 109
  1. Solar heating.   2. Solar houses—Design and
construction.   3. Greenhouses—Design and construction.   I. Title.
TH7413.K7   697'.78   74-34332
ISBN 0-684-14200-7

This book published simultaneously in the
United States of America and in Canada—
Copyright under the Berne Convention

No part of this book may be reproduced in any form
without the permission of Charles Scribner's Sons.

1 3 5 7 9 11 13 15 17 19 MD/C 20 18 16 14 12 10 8 6 4 2

Printed in the United States of America

Acknowledgments

Several of my friends helped me in the course of writing this book. Some were designers and architects, others carpenters and builders. Other help came from folks I met along the way who were also interested in sun-heated rooms and by chance were trying to design one for themselves. To all these people, my thanks, and I want to express special gratitude to Andrew Roy Addkison, Environmental Designer, at the California College of Arts and Crafts in Oakland, California, who contributed freely of his knowledge of building and design. To Adrian Martinez, I owe a great deal of thanks for interpreting some of my design ideas for the various rooms illustrated in this book. Also, I wish to note my appreciation of my editor, Elinor Parker, whose enthusiasm and advice made this a better book.

JACK KRAMER

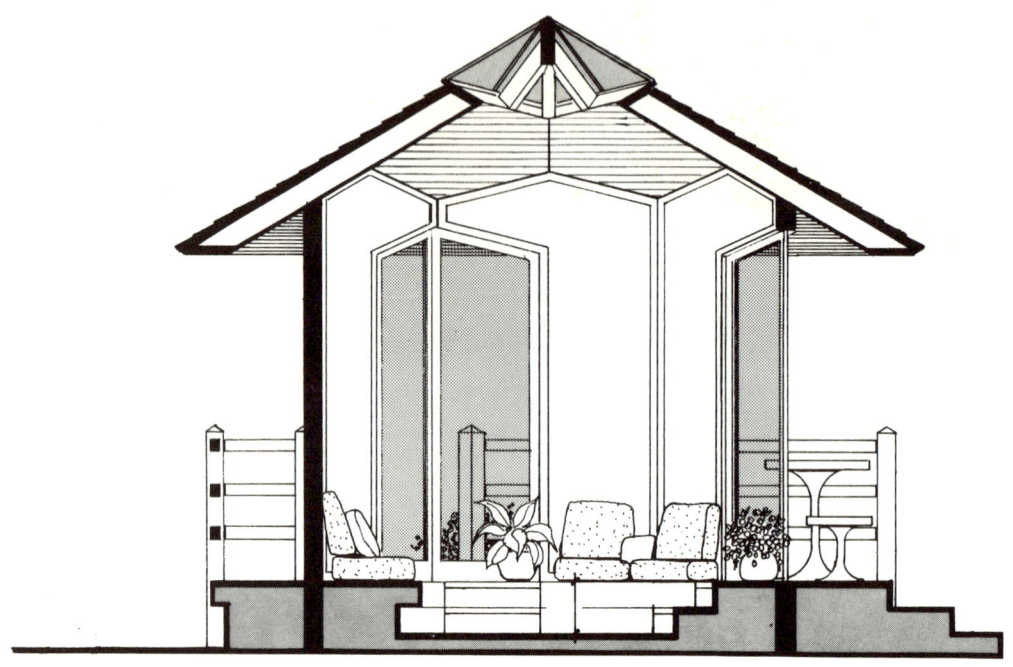

Author's Note

I think this book has always been in my mind but it took some words from my editor, Elinor Parker, to bring it to fruition. She told me of her various visits to sun-heated summerhouses when she was last in England. That was the key to this book.

Contents

Introduction: Stretching the Year   xi

Part One: THE INDOOR/OUTDOOR ROOM

1   The Indoor/Outdoor Room   3
      The Garden Room
      The Solarium
      The Gazebo
      The Summerhouse
      My Indoor/Outdoor Room

2   Using the Sun's Heat   20
      Solar Structures
      How It Can Work For Us

3 Your Personal Place   30
    Basic Building Considerations
    Shape and Design
        Professional Design Help
    Materials

4 Building the Indoor/Outdoor Room   37
    Footings
    Concrete Floors
    Brick Floors
    Walls
        Laying Block
        Brick Wall
        Concrete Wall
    Glass and Glazing
    Roofing
        Domes

5 Making the Indoor/Outdoor Room Comfortable   55
    Benches, Furniture
    Plant Furniture
    Super-plants
    Suitable Conditions

Part Two: THE GREENHOUSE

6 A Place For Plants   65
    Pit Greenhouse
    The Greenhouse (Lean-to)
    The Geodesic Dome
    Placement and Size
    How to Get Started

7 Construction Know-How   75
    Pit Greenhouse
    Lean-to Greenhouse
    Dome Greenhouse
    Foundations

8   The Greenhouse For You   87
        For Temperate Climates
        For Cold Climates
        Site
        Planting
        List of Trees
        List of Shrubs

9   Planning and Planting   101
        Benches, Shelves, Hangers
        Hardware Accessories
        Shading
        Problems

Bibliography   109

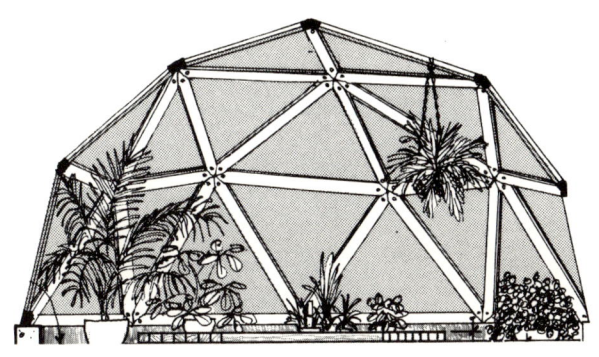

Introduction: Stretching the Year

I HAVE NEVER HAD A CONVENTIONAL GREENHOUSE; INSTEAD I OPTED to build my own (what I call indoor/outdoor rooms), and have designed several through the years—one in Illinois, four in California where I now live. My preference for this kind of structure is that it can be a place for people as well as for plants. I use my indoor/outdoor room in the chilly but sunny days of early spring and late autumn where during the day (with no artificial heat at all) the temperature is about 70°F and at night about 55°F (not too cold to enjoy).

Essentially a sun trap, indoor/outdoor rooms whether you call them garden rooms, summerhouses or "greenhouses" are really worth their space in golden hours of enjoyment, not to mention the savings in fuel bills. In summer, of course, rooms can be very hot but proper screening (roller blinds, trellises) help to make the area comfortable.

In a similar vein and using the same sun-heated idea, a partially underground greenhouse can be built specifically for plants. The Egyptians used such structures and so did the North American Indians to grow plants. In very cold climates the glass areas of these structures can be protected from weather with panels of wood or insulating materials to conserve heat.

This book shows several ways to best use the energy of the sun as a source of heat for indoor/outdoor rooms (whether for people or for plants). To function properly, the rooms must be placed strategically on the property to benefit from the sun, be built with overhangs and materials that use nature to the fullest to create a self-sustaining area.

In any case, you will find that whether you decide to have a garden room, a summerhouse or a pit greenhouse or a lean-to, you will find information and help within these pages to help you save money and to build the best possible indoor/outdoor room for you. My concept is, and I have found it true, that these rooms stretch the season and except in very cold climates will be usable from late February to late November rather than only a few months in summer.

JACK KRAMER

Part One: The Indoor/ Outdoor Room

1 The Indoor/Outdoor Room

BRINGING THE OUTDOORS INDOORS HAS BEEN A POPULAR PASTIME in the last twenty years. Even in severe winter climates, homeowners more than ever are using some of their outdoor areas (small though they may be) as enclosed or partially enclosed total living space. (Apartment dwellers too have brought greenery into their living day with pot plants, making the rooms resemble greenhouses.)

Just what form the sun-heated indoor/outdoor room takes and where to put it are people's dilemmas. More used for people than plants, this indoor/outdoor space may be called a (1) garden room, (2) solarium, (3) gazebo, or (4) summerhouse. It may be a new addition, a reconverted porch, or a modified prefabricated greenhouse. It may be attached to the house or a separate building.

If properly planned and designed, and built with proper materials, the room can function inexpensively as an additional space to greatly increase your love affair with nature. So no matter what you call the room or where it is, it should be constructed with materials to help retain heat, and it should be south-oriented to capture as much of the sun's rays as possible. The room should also be protected by natural barriers—hedges, shrubs—to thwart weather. In essence, you want a sun-heated, self-sustaining indoor/outdoor room.

The Garden Room

This indoor/outdoor area may also be called a plant room or Florida room. It is an enclosed structure generally attached to the house and easily adjacent to it. It is a place for morning coffee, relaxing during the day, and occasional entertaining. The room has some glass in its construction, but it is not a greenhouse nor is it predominantly glass. Indeed, such construction would defeat its purpose as a sun-heated room because glass is a poor insulator; it would be too hot in summer and too cold in winter. The garden room has some glass in the ceiling (facing south) and some glass in its wall construction. If built with proper heat-retaining material (explained later), it is a space that can be used from early morning to dusk from March to November. Only in winter would auxiliary heating be necessary. It stretches the seasons and as such is a valuable asset to the home. While it is predominantly a place for people it should have some plants to create a total comfortable environment.

This garden room adds great beauty to the house and while it is a place for people to relax and enjoy conversation, it is also a place for some plants. The room is situated so that the hill in the rear acts as a buffer for strong winds, the floor is brick to retain heat and four panels of glass and four panels of insulated board (covered with fabric) are used for a ceiling. The room faces south. (Photo by Joyce R. Wilson).

Section

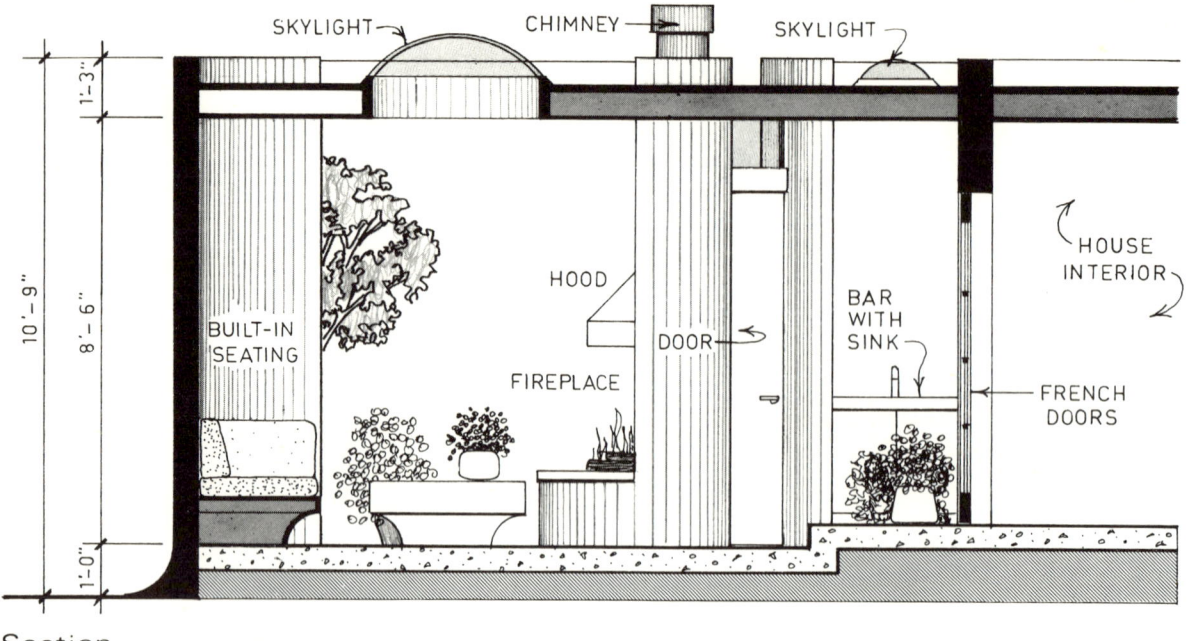

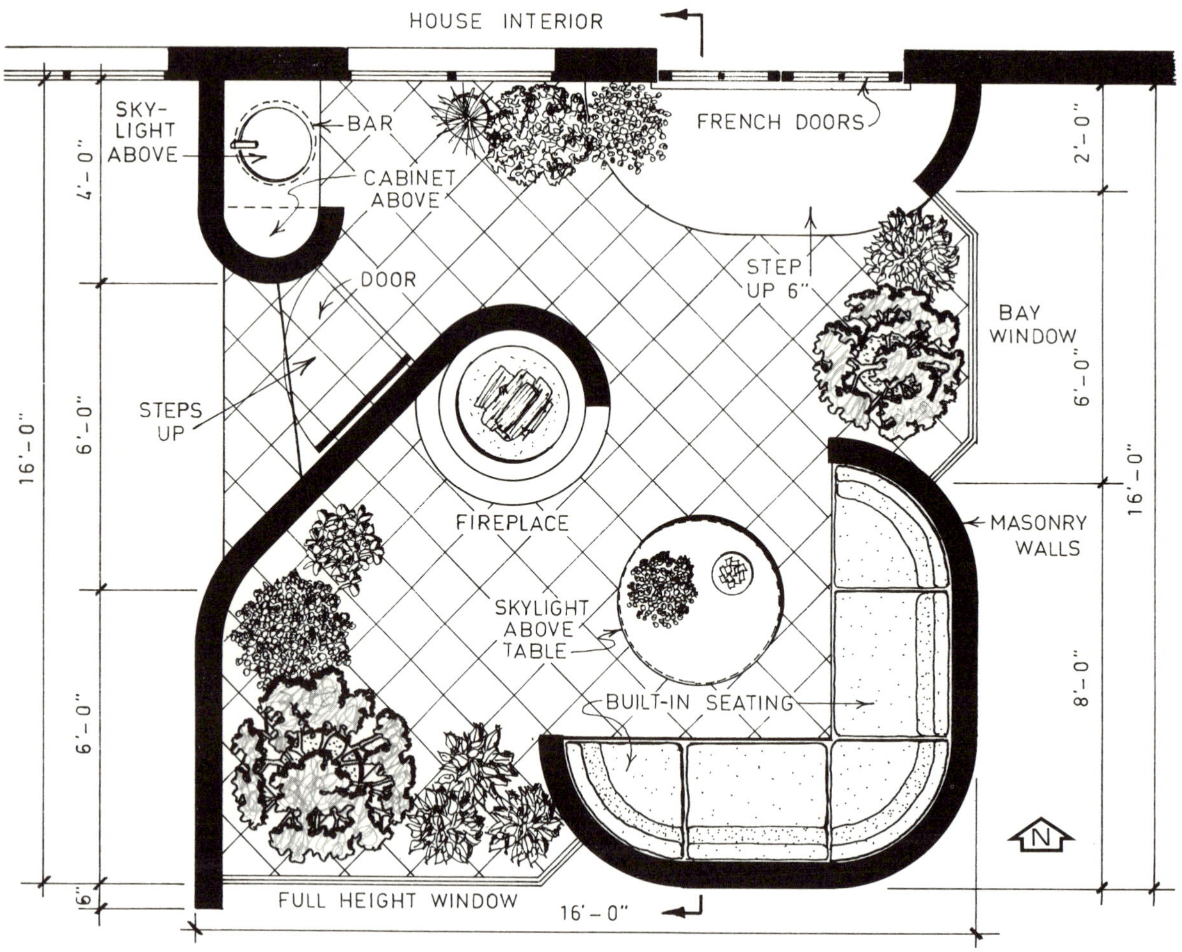

Attached Garden Room

Adrian Martinez

Sunken below the level of the house and protected with a concrete wall this sun-heated garden room requires little artificial heat. The room can be used almost all year and is a comfortable place for people as well as plants. (Photo by Matthew Barr).

The Solarium

I saw my first solarium—an octagonal glassed-in small room—when I was visiting an old house in Chicago. It adjoined the living room, closed off with leaded-glass doors, and was indeed a handsome sight. When I went into the solarium to look at my friend's plants I noticed how hot it was and at the same time I thought how terribly cold it would be in the evening because of the total glass construction.

The solarium of Europe in the 16th century was generally a place for elderly people or the ill to rest during the day in warmth and in good humidity. Plants, of course, were part of the room and added to its beauty. Certainly the solarium has its uses but for our purposes it is an expensive room to heat on cold nights and a costly and difficult room to build. While it is a sun-heated room it can be used just a few short months and then only when the sun shines. In winter the solarium would require a great deal of artificial heating. (Or else could not be used at all for people or plants with the exception of winter-resting plants.) For our purposes the solarium is the least desirable (moneywise) of the other structures.

A solarium in England shows the heavy construction using masonry to retain heat. The windows face south to catch as much sun as possible. Such grand structures hardly apply to today's scheme of living but the basic concepts of construction—to retain heat—are quite valid. (Photo by Roger Scharmer).

The Gazebo

The gazebo is generally an open building but can be enclosed and used as a green retreat. This room, like the summerhouse that follows, was a decorative addition to many old English estates. Ironically, today, even on small properties the gazebo is reappearing either as a poolside structure or simply as an enclosed extra outdoor room.

Generally, the gazebo is separate from the house and as mentioned may be used as a poolside place, or simply a place to relax or to grow plants or to provide a pleasant vista for the viewer. The gazebo is unique because in terms of design it can assume many shapes and thus add beauty to the property while acting as a decorative and functional unit. The old-fashioned gazebo was small, charming and intimate; never spacious and sprawling.

Sun-heated, the gazebo offers several advantages. It is inexpensive to build, can be used a great deal of the year (if enclosed and built properly) and is a departure from the usual outdoor room.

Recently, manufacturers realizing the value of gazebos are offering knocked-down kits with all components for open gazebos. These can be modified to make them enclosed havens.

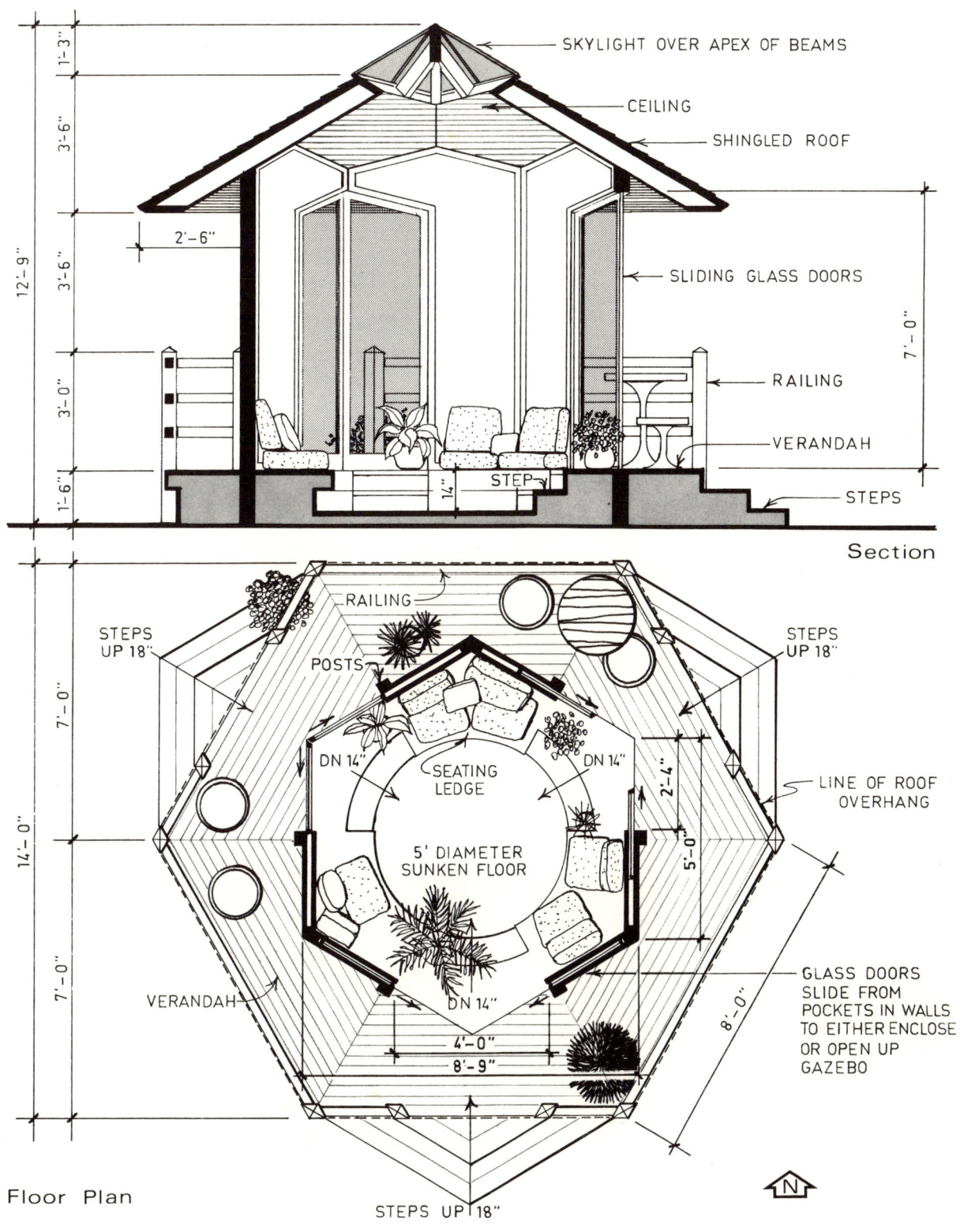

Section

Floor Plan

Indoor/Outdoor Gazebo Adrian Martinez

(*Above*) This gazebo in a temperate-all-year climate is both for people and plants. The arched construction is simple and there is great beauty here. (Photo by Matthew Barr).

(*Opposite page*) The gazebo is a detached structure and this one is used solely for people; it is a pleasant retreat away from the house during the summer months. Enclosed, (a different design), it could also function for winter-resting plants. (Photo courtesy California Redwood Association).

The Summerhouse

The summerhouse, a detached structure especially popular in England in the 1860s, also has many possibilities as an indoor/outdoor space. It lengthens the seasons in two directions (spring and fall) and provides more living area. This structure may accommodate some plants, but usually it is a place for people to view nature (like the gazebo), to read or relax away from the house, with a few plants here and there for decoration. Originally built of natural materials—stone, brick, earth—and unheated, the summerhouse used the warmth of the sun as the chief heating source. These structures were highly ornate; some had pillars or posts and were Chinese or Victorian in character. Whether open or closed (or closed in center, open at sides), most had seating benches. The summerhouse can also be attached to the house, although in this case the design of course would be quite different (matching the character of the house) from the separate unit.

A typical English summerhouse of heavy construction using masonry and glass. It is ideally located to provide seclusion for the users. (Photo by Roger Scharmer).

My Indoor/Outdoor Room

My place for plants and people takes a little from the garden room and solarium and much from the summerhouse construction. It depends entirely upon the sun for its heat and faces south, with the north wall as the common house wall and the east and west walls containing a small portion of glass. It works with nature, not against her. On cloudy days it is of course chillier than I would like, but it is still suitable for both plants and people; in early evening you might need a sweater. Because of high heat buildup during the day (the south wall radiates great heat) I have louvers and windows that can be opened easily to allow air to flow through the room and keep it comfortable even on the hottest days.

Very little glass is used in the ceiling, and building materials are redwood for walls, thick hemlock for ceilings, and concrete piers with a brick floor. I use this room 10 months a year from early morning until evening. In January and February I use it during the day; in the evening it is chilly (about 50°F). The room is closed off from the main house (living room) by French doors.

No matter what you call your indoor/outdoor space, or whether it is attached to the house or separate, remember that it should be (1) built with heat-retaining materials, (2) placed to capture as much of the sun's heat as possible, and (3) geographically situated to benefit from natural vegetation. If so, it will be a room inexpensive to heat and a room in which you will find yourself spending more and more time in all but the coldest months of the year.

The author's sun-heated indoor/outdoor room. The floor is concrete to retain the sun's heat. Great expanses of double-glazed glass that face south capture heat and the ceiling is 3/4 inch hemlock. The seating arrangement at right accommodates six people. This room is used from early spring to late fall; there is no artificial heat. (Photo by author).

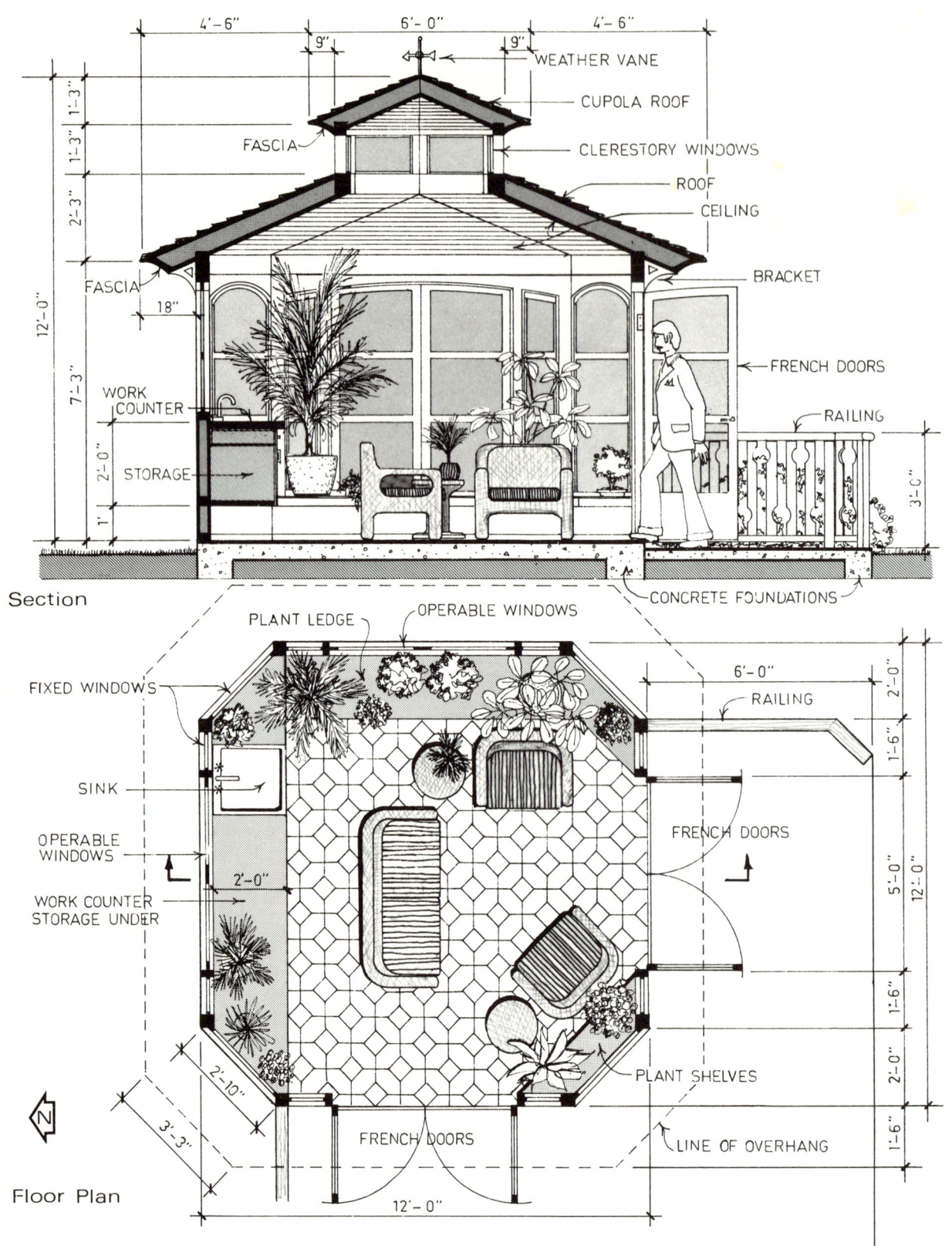

Indoor/Outdoor Room

Adrian Martinez

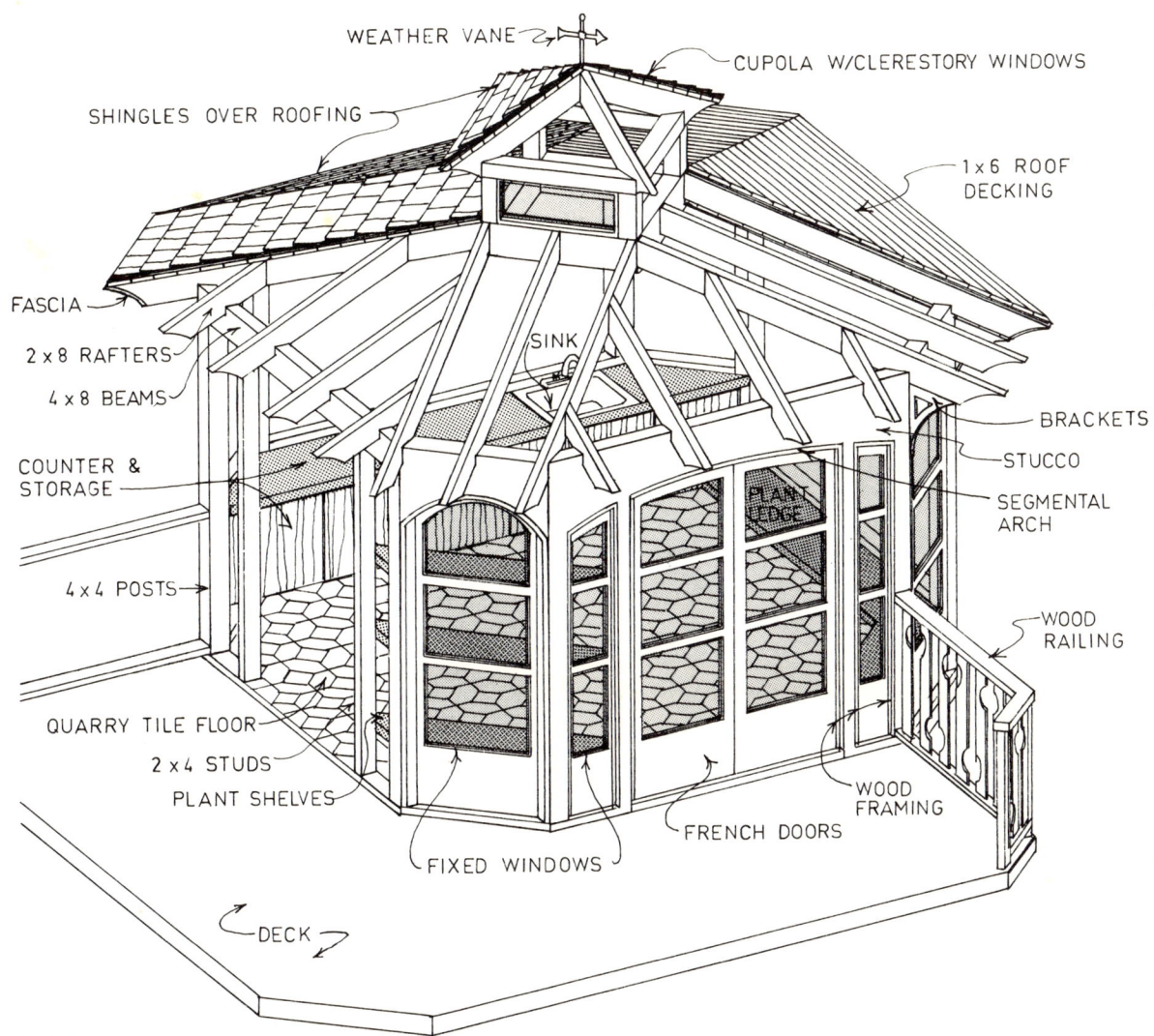

ROOF
1 x 6 ROOF DECKING, ROOFING, REDWOOD SHINGLES
2 x 8 RAFTERS W/INSULATION IN BETWEEN, CURVED
PLYWOOD FASCIA, CUT-OUT WOOD BRACKETS

CUPOLA
4 x 4 FRAME, 2 x 8 RAFTERS, FIXED WOOD FRAMED
CLERESTORY WINDOWS

CEILING
PAINTED 1 x 6 T & G BOARDS, INCLUDING CUPOLA &
OVERHANGS

WALL STRUCTURE
WOOD FRAME, 4 x 4 POSTS, 2 x 4 STUDS & FRAMING,
PLYWOOD SHEATHING, METAL LATH, STUCCO EXTERIOR,
PLASTER INTERIOR

WINDOWS
WOOD FRAME & GLASS, CORNER & END WINDOWS
FIXED, OTHERS OPERABLE

FLOOR
QUARRY TILE SURFACED CONCRETE, REINFORCED
CONCRETE FOUNDATIONS

DECK
WOOD DECKING OR QUARRY TILE OVER CONCRETE

RAILING
PAINTED WOOD, ROUNDED 2 x 4 POSTS & RAILS W/
CUT-OUT 1 x 6's

COUNTER & STORAGE
WATERPROOFED LAMINATED WOOD TOP W/STAINLESS
SINK, PLYWOOD STORAGE CABINET

PLANT LEDGE & SHELVES
WATERPROOFED LAMINATED WOOD

DOORS
DOUBLE FRENCH DOORS, WOOD FRAME & GLASS,
TOP CURVED TO MATCH ADJACENT WINDOWS IN
A SEGMENTAL ARCH

Indoor/Outdoor Room

2 Using the Sun's Heat

ALL THE INDOOR/OUTDOOR ROOMS WE DISCUSS HAVE ONE THING in common: they use the sun as a source of heat. Solar heating, however, encompasses a gamut of ideas. At the University of Delaware, a house called Solar One converts sunlight into both electricity and heat for home use. The home is designed to get up to 80 percent of its heating, cooling, and electrical needs from the sun. Other research projects are in experimental stages at various places in the United States, each with somewhat different methods, but all having one thing in common: getting heat from the sun and converting it into energy, in most cases without mechanical

The famous orangerie or greenhouses at Chatsworth show a lesson in sound building construction for any indoor/outdoor room that wants to use nature to supply heat. The base and walls of the structures are heavy masonry and thus retain heat a long time. The real walls (opposite glass walls) are gray—a dark color—to retain heat. Overhangs of decorative grillework and slightly canted skylights complete the handsome structure. (Photo by Roger Scharmer).

A close-up view of the orangerie at Chatsworth. (Photo by Roger Scharmer).

aids. Harold Hay at Atascadero, California, uses insulating panels over water bags on the roof. Other experimental houses have glass panes over part of an aluminum roof. And solar water heaters that were used in Florida in the 1930s are once again being manufactured.

When I was a boy in Florida in the 1940s, our house was called a solar house because the sun heated water for our baths and dishes. A rooftop solar water heater was used. The principle was simple. This boxlike unit was made of heat-absorbent material backed with heavy insulation and covered with glass. The box was on the south eave of the roof, and the glass panels covered the "collector" to keep heat from reflecting back. The water was pumped up to the space between the panels, heated, and returned to an insulated (holding) tank for storage. Needless to say, on cloudy days we had little hot water, so in a few years my family installed an inexpensive auxiliary system to help heat the water. This system still works today.

Solar Structures

There are many ways to solar heat, but perhaps the simplest is to have a south-oriented structure. Essentially, a house that faces south allows the sun's rays in winter to penetrate through windows into the house and provide considerable warmth. In summer the path of the sun is overhead and above the roof, so the structure is shaded. The house is built so the south side has glass and is higher than the north side, which is low to keep out cold winds. This is not strictly a solar house, but a large window design on the south side of a house with adequate overhang does work, provided there is no excessive heat loss through the glass (thermopane or double-glazed glass is used). I lived in such a house (built in 1949) in Chicago for two years in the late 1960s. The north side of the house had small windows and heavy stucco walls; the west side, like the south side, had more glass. In cold climates there is only about 8 to 10 hours of daylight, and if the sun shines every other day, you have sunlight only a portion of the time. (I am basing this upon my years in Chicago.)

Other solar structures have different concepts, but the principle remains the same. At the University of Colorado, a 12- x 20-foot trap was put on the top of a house. Air entered the trap at the eaves; passed through

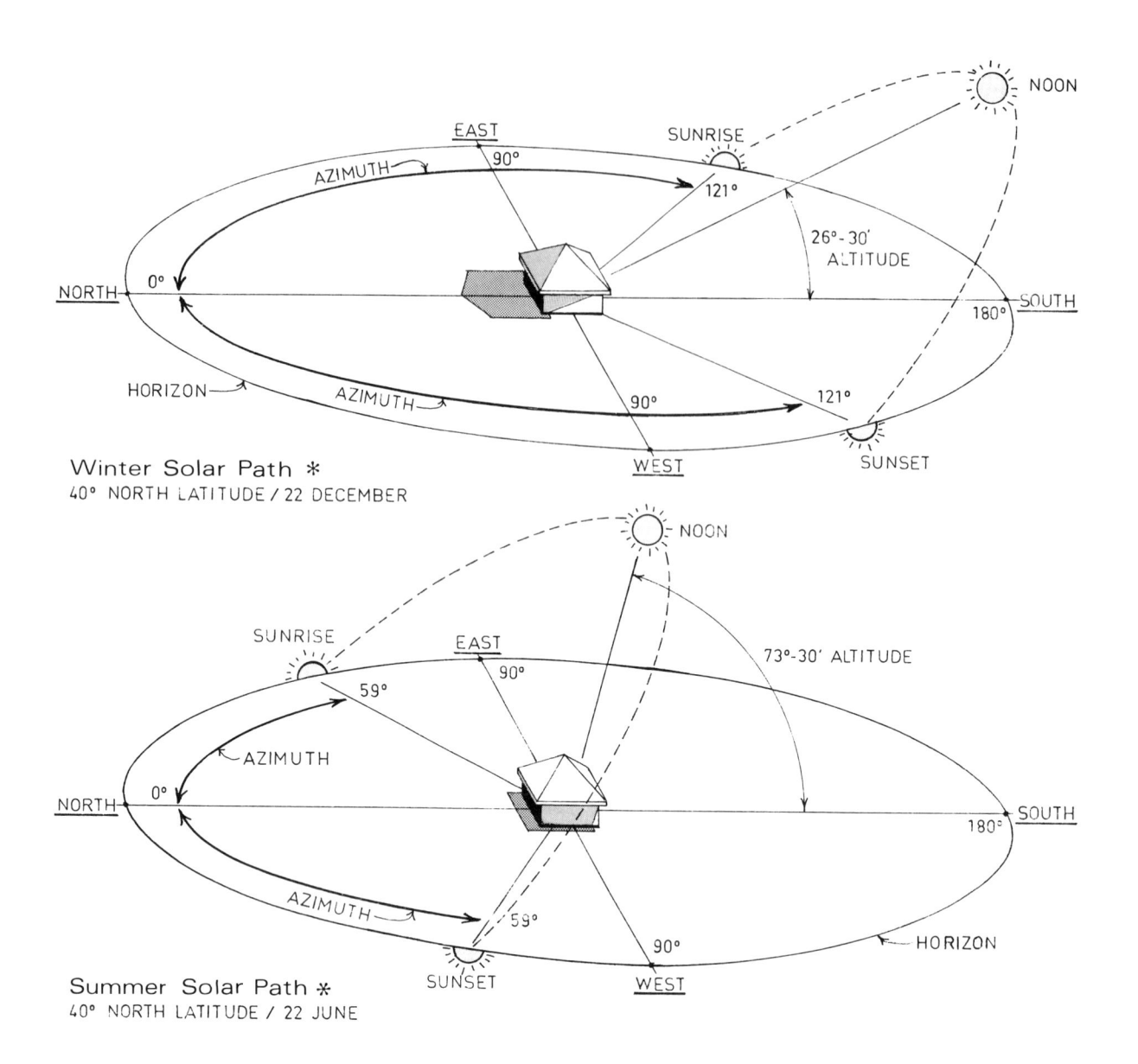

Winter Solar Path *
40° NORTH LATITUDE / 22 DECEMBER

Summer Solar Path *
40° NORTH LATITUDE / 22 JUNE

LATITUDE	SEASON	SUNRISE	SUNSET	AZIMUTH	ALTITUDE
50°	WINTER	8:00	4:00	128°-30'	16°-30'
	SUMMER	4:00	8:00	51°-30'	63°-30'
45°	WINTER	7:40	4:20	124°-30'	21°-30'
	SUMMER	4:20	7:40	55°-30'	68°-30'
* 40°	WINTER	7:30	4:30	121°-0'	26°-30'
	SUMMER	4:30	7:30	59°-0'	73°-30'

LATITUDE	SEASON	SUNRISE	SUNSET	AZIMUTH	ALTITUDE
35°	WINTER	7:10	4:50	119°-0'	31°-30'
	SUMMER	4:50	7:10	61°-30'	78°-30'
30°	WINTER	7:00	5:00	117°-30'	36°-30'
	SUMMER	5:00	7:00	62°-30'	83°-30'
25°	WINTER	6:50	5:10	116°-30'	41°-30'
	SUMMER	5:10	6:50	88°-30'	63°-30'

NOTE: THESE LATITUDES COVER THE CONTINENTAL UNITED STATES. HOURS INDICATED ARE STANDARD TIME. AZIMUTHS ARE AT SUNRISE AND SUNSET, NOON AZIMUTHS ARE ALWAYS 180°. NOON ALTITUDES ARE GIVEN, ALTITUDES AT SUNRISE AND SUNSET ARE ALWAYS 0°.

Solar Angles

a double-glass panel, with the bottom layer painted black to absorb heat; and was carried by ducts to the furnace-pipe system. By controlling the speed of the air movement through the layers of glass you could control the temperature.

Other houses do not use a collector but rather—like primitive North American Indian houses—may have a concrete slab that forms the floor. The floor stores the heat that comes in from the windows. The slab is 6 inches thick, poured over washed gravel and insulating board. A concrete floor slab is a remarkable heat storer: it absorbs enough heat during the day (provided the sun shines) to keep the room warm throughout the night in all but a few cold weeks of winter. Black-surfaced masonry walls in back of the south glass wall will also absorb sun, and thus heat. The heat is trapped in the walls and at night warms the room.

How It Can Work for Us

The examples just given were standard residential houses with approximately 1000 square feet. For our indoor/outdoor room we are talking about much less space, say, 200 square feet, or a room about 10 x 20 feet. This makes trapping the sun's energy and conserving it for heat much less complex, although the same principles apply. In addition, the building can also be partially buried in the ground (earth is a fine insulating material), with the west and south window areas open. At night, shutters or heavy drapery can be used as an insulating cover. This is an easy and inexpensive way to increase heat conservation.

For your indoor/outdoor room you can try converting commercial collectors: the heat is collected and stored, and the box collector uses air or water circulation to carry heat from it to a holding tank; the hot water then is run to pipes through the walls of the indoor/outdoor room. Another way to use the collector is below the floor level; then gravity-flow, hot-air circulation is used via ducts buried in stone below the masonry floor. The glass collector is tiled to absorb as much of the winter sun as possible and reflect the high-angle summer sun's rays. (For more information about solar collectors, write to W. A. Robbins & Son, 1420 NW 20th St., Miami, FL 33142, or Sky Therm Processes and Engineering, 2424 Wilshire Blvd., Los Angeles, CA 90067, Attention Harold Hay.)

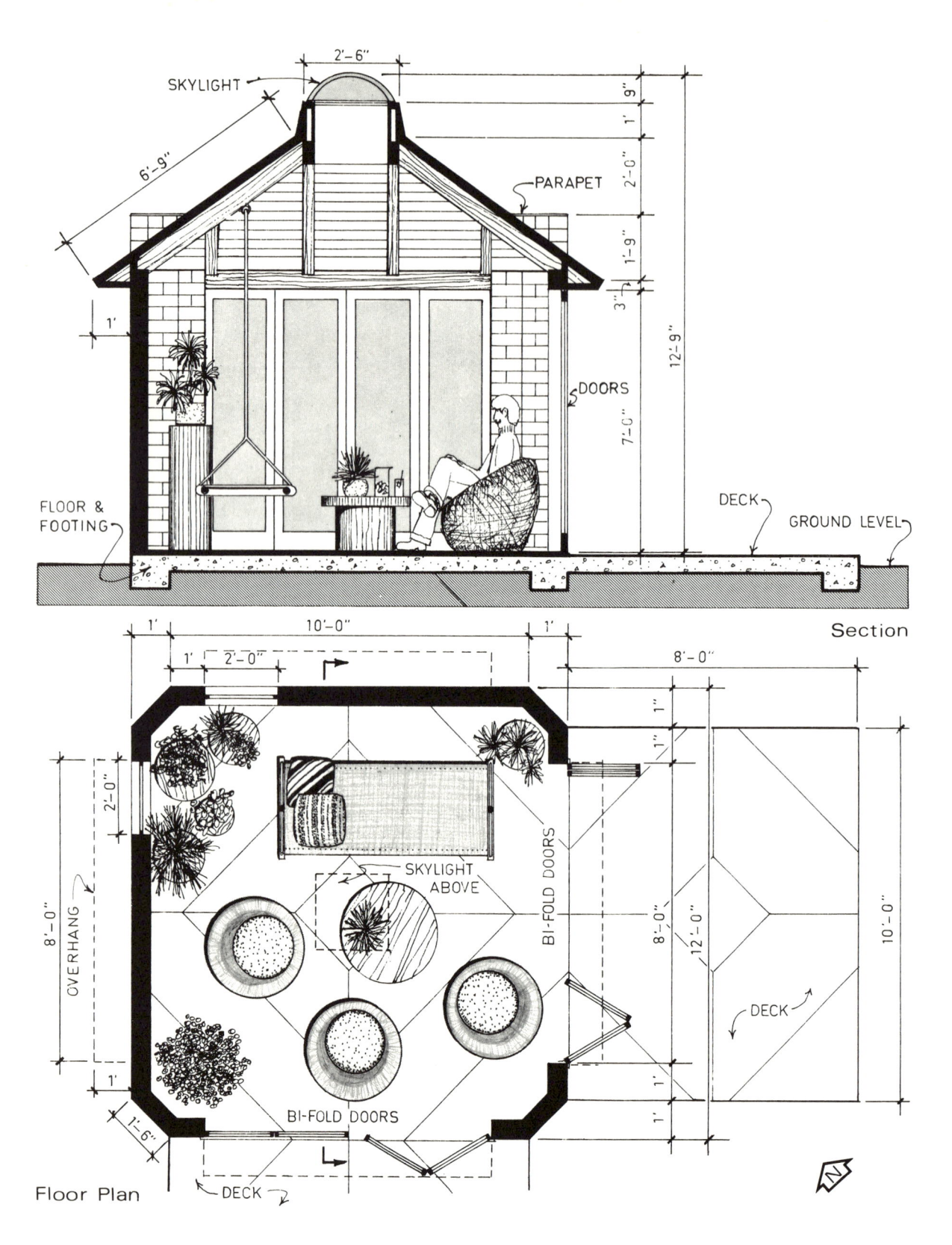

Masonry Summerhouse

Adrian Martinez

Section

Floor Plan

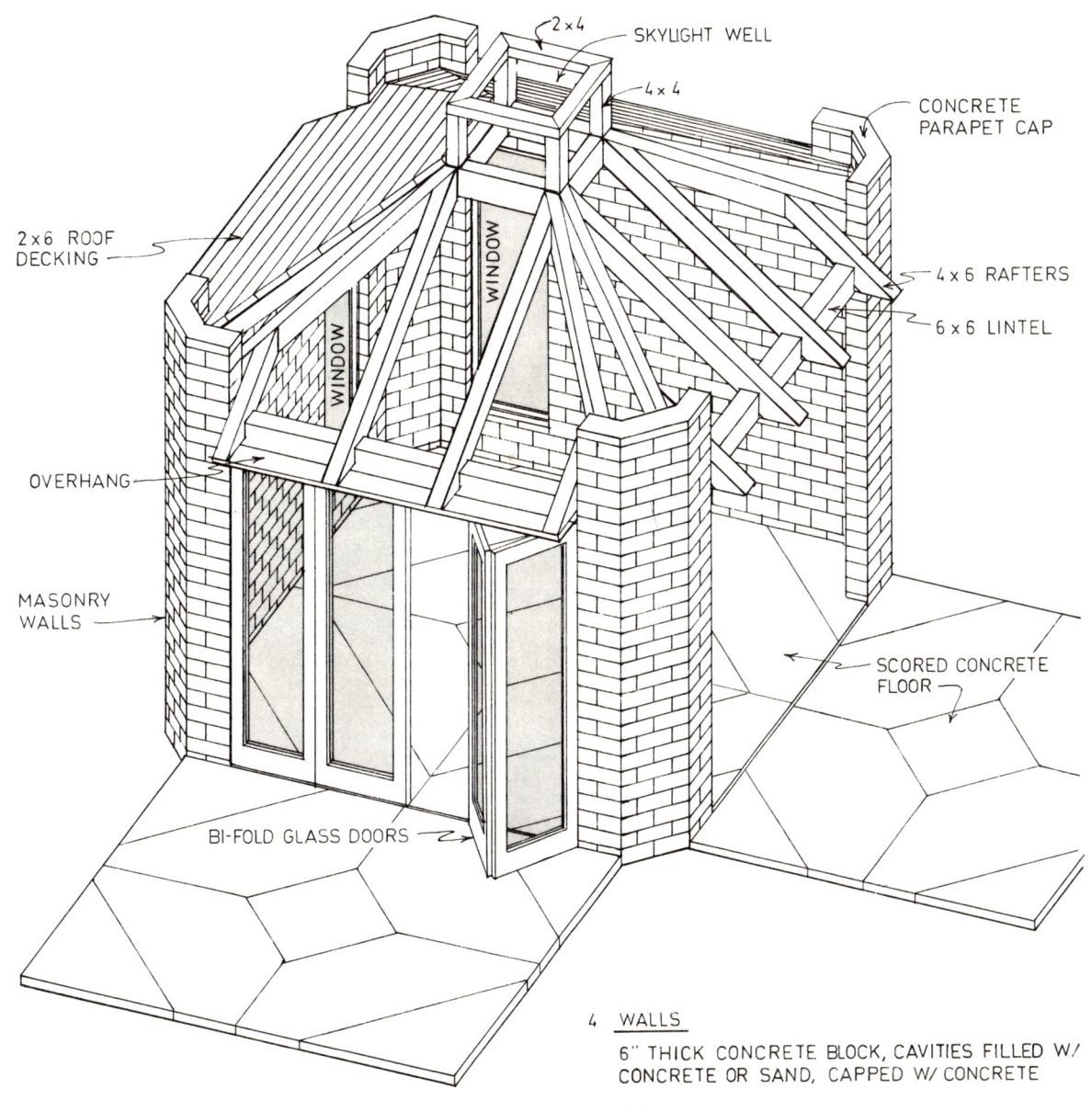

1 SKYLIGHT

ACRYLIC DOME 2'-6" SQUARE BASE, 9" HIGH

2 SKYLIGHT WELL

WOOD FRAME 4-4x4's x 12" & 4-2x4's x 2'-6"
PLYWOOD CLAD, PAINTED INSIDE, SHINGLED OUTSIDE

3 ROOF

WOOD RAFTERS 16-4x6's x 7' & 8-4x6's x 3'
COLLAR AT TOP 4-4x6's x 2'-6"
LINTELS OVER DOORS 2-6x6's x 8'
2x6 ROOF DECKING W/BUILT-UP ROOFING & SHINGLES
4-1x12's x 8' UNDER OVERHANGS

4 WALLS

6" THICK CONCRETE BLOCK, CAVITIES FILLED W/
CONCRETE OR SAND, CAPPED W/ CONCRETE

5 FLOOR

REINFORCED CONCRETE, SCORED, EXPOSED
AGGREGATE SURFACE

6 WINDOWS

FIXED, WOOD FRAMED GLASS 2-2'x6'

7 DOORS

BI-FOLDING, WOOD FRAMED GLASS 8-2'x7'

NOTE: TO BEST UTILIZE SOLAR HEAT, WALLS &
FLOOR SHOULD BE DARK, GLAZING SHOULD BE
DOUBLE INCLUDING SKYLIGHT

Masonry Summerhouse

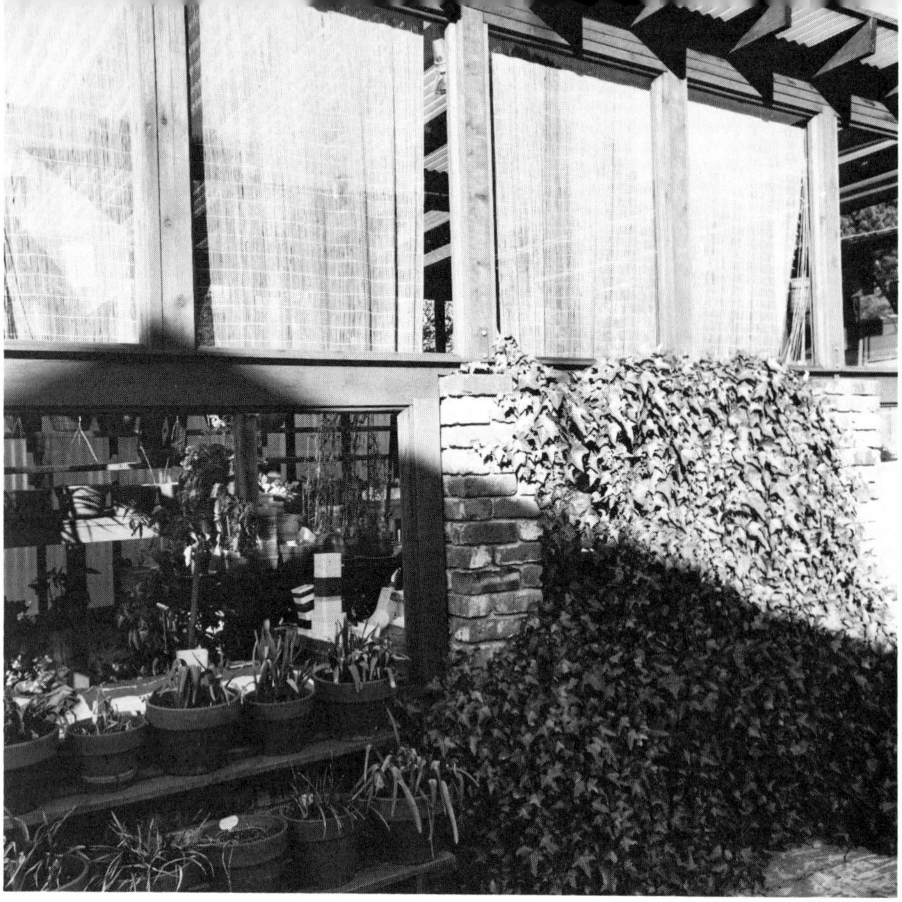

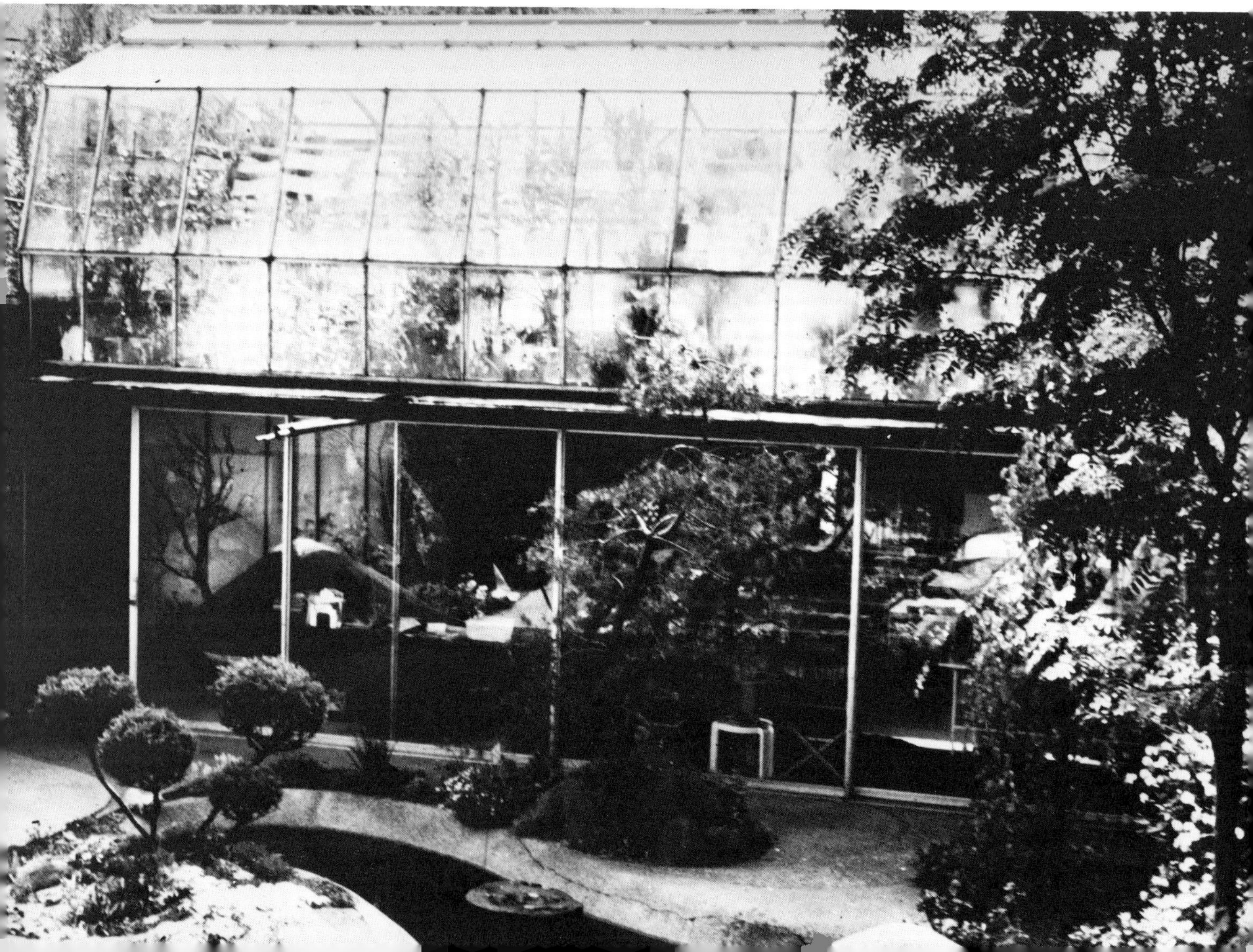

Because the indoor/outdoor room is generally small—you want it intimate to be charming—it is well suited to solar-heat methods. Whether the house is sunken or double glazed with thermopane, or whether you use solar collectors or masonry walls and ceilings, the additional cost decreases heating bills. Essentially this is a structure that heats itself and can be used daily by you except during very cold winters.

Just how you adapt the many kinds of solar principles to your retreat depends upon your own climate and uses for the building. (See Chapter 8.) But the point is, and it is a big one, this kind of indoor/outdoor room can stretch your year by months. Actually, if you wanted to use it in winter, as mentioned, suitable shutters could be fitted in place at night so it would be livable by day even in the coldest weather. Of course, on days without sun it could not be used (unless you have a gravity pipe system to circulate the hot water), but remember that without any solar innovations at all you cannot use the room in most climates from November until May.

(*Opposite top*) This sunken structure for plants and people is well designed with brick as walls—again to retain heat. Being below grade the room is naturally warmer than if at ground level. Such design cuts down on artificial heat and makes the room a livable place in all but the coldest months of the year. (Photo by Matthew Barr.)

(*Opposite bottom*) This solar greenhouse uses small panes of glass to capture as much of the sun's heat as possible. The design of the roof is unusual and attractive. The floor is concrete aggregate and the north, east and west sides are protected by walls so strong winter winds are buffered. (Photo by Pat Matsomoto).

3 Your Personal Place

THE SUN-HEATED ROOM IS CHEERFUL AND WARM, JUST THE WAY your kitchen is when sunlight streams through the windows. A separate room strategically placed and with some (but not all) windows will capture the sun and its warmth and thus be a comfortable blessing on a chilly autumn morning or a cold winter afternoon. But most of us do not have or cannot afford a custom-designed house with a sun-heated room, so we must add to the existing structure, with either a detached summerhouse in the garden or a room attached to the main house.

Whether your indoor/outdoor room is to be attached or separate, certain placement, design, and materials details have to be considered. By adhering to this chapter's "rules," you not only add to the functional ability of the additional space, you enhance the beauty of the main house. (See the next chapter for detailed construction and materials facts.)

The design of this summerhouse is simple and yet attractive. Its remote position has its purpose—to provide a quiet retreat. It is on a slight knoll to afford a lovely view of the brook. (Photo by Roger Scharmer).

Basic Building Considerations

Ideally, any type of indoor/outdoor room should face south, to get as much of the sun's warmth as possible. If possible, locate the structure (if it is detached) on a higher piece of ground than the main house. This adds dimension and some drama, and elevated, the structure will command a better view. (A slight protected knoll or hill is perfect.)

You should also plan on working with rather than against nature, which means utilizing natural barriers and thinking of climate. For example, trees and shrubs will bear the brunt of prevailing storms and winds in the winter and absorb the heat in summer. Thus, in cold regions you would protect the east side of the room from winds by using hedges of evergreen shrubs. Some natural barriers placed a distance of, say, 50 feet from the north wall will allow summer breezes to circulate around the structure for cooling. And of course you should not obstruct the south wall because you want to catch the sun's rays for warmth. Nature will not only serve you; she will also offer perspective and make your room part of the whole landscape rather than a tacked-on afterthought. As previously mentioned, use materials in the building of the room that will help retain heat, such as concrete or earth floors, masonry north walls, and glass south walls.

Shape and Design

A plain square or rectangular box placed near or attached to the main house will look just like that—a box. Use good linear lines and design to make a handsome addition to the property and to make it visually inviting. The indoor/outdoor room can be hexagonally shaped, or try a more elaborate octagonal structure or dome. The character of the room can be Chinese, Grecian, or very rustic. But no matter what its shape or character, the room should be architecturally pleasing; it will not cost any more than an unappealing room.

As mentioned in Chapter 1, the added living space may be a separate building away from the main house (summerhouse), or it can be attached to the house as a lean-to (garden room), utilizing one wall of the house as

This indoor/outdoor room has a greenhouse quality and yet is not stereotyped. It is attached to the house and is a pleasant place to sit amidst plants in early spring or late fall. (Photo by Matthew Barr).

a side, which saves some money. Each design has its advantages. The unattached structure takes you out of the house and provides a complete change of scenery close to the garden. Its detachment also puts you out of the reach of telephones and other daily distractions. Its disadvantage is that in inclement weather you must face the summer winds or rain to get to it. The attached unit (off the living room, kitchen, or wherever) has easy access and becomes part of the house, thus expanding your indoor area—an extra room. Its disadvantage is that it is still part of the home and offers little change in scenery, or closeness or proximation to nature. I suppose two small units, one attached to the home, the other separate, would be ideal, but this is usually beyond the average person's pocketbook.

Be careful when you design the enclosed space *not* to plan on its being all glass, or it will be too hot in summer and too cold in winter. A good idea is to make the south wall at least 50 percent glass, with the ceiling not more than 30 percent glass.

Any design should also consider shading and ventilation facilities. Roll-up blinds may be necessary in the hottest months to keep the indoor/outdoor room reasonably comfortable; overhangs and screens can also combat devastating heat rays. I find that jalousie windows at either end of the room are good for ventilation; they can be easily opened or closed, thus furnishing adequate air circulation for plants and people. (Clerestory windows are also good ventilation ideas.)

You can seek professional design help (see next section), but if you are doing your own design, draw rough sketches that you can show to a contractor. (You can, of course, construct the room yourself, following the drawings in this book, but you will need final drawings by an architect for city approval and building permits.) To help with your design, send for the various brochures from lumber associations for garden buildings, gazebos, and such; the plans and drawings can be converted to your design. You may have to add more glass, take away some, redesign, and so forth, but at least to start you will have some grasp of an outdoor structure. Use what you need, and discard other ideas to finalize your own indoor/outdoor retreat.

Professional Design Help

It may be difficult to find architects and designers familiar with indoor/outdoor room construction—most professional people cling to the idea that you are talking about a greenhouse. But this is precisely what you do not want. Explain that the room will be used for people more than for plants and that it should be constructed to save fuel rather than add to the cost of heating. Tell your consultant that the room should face south, should be made of natural materials, and should be small or medium-sized (large areas are hard to heat and not as inviting). Finally, do not forget to discuss roof design (this is vital). You may want to consider the many plastic dome skylights as well as the conventional glass ones.

You will need some rough sketches to get started, but a complete set of drawing plans is not necessary at this design stage. The consultant will advise you about the best location on the property, materials, and cost.

Materials

Utilize as many natural materials as possible, for example, wood and masonry (brick, concrete block). Wood is generally the basic building material, so learn something about the various grades and kinds of lumber. Glass is another important material; there are many kinds. Because you are trying to use as much of the sun as possible, selecting the right type of glass to admit the most sun rays and getting glass sealed properly in walls and ceilings is vital. You must have airtight glazing and no leaks in any areas or you will have a heat loss of 30 to 40 percent. Use thermopane glass (also called double-glazed) or 1/4-inch plexiglass.

If the room is to be of a masonry construction, know something about concrete and brick: how it is put together and just what it will or will not do. A concrete or brick floor will conserve heat that accumulates during the day and will not be stained by water. Concrete can be poured, or you can use the many different types of concrete blocks (sold by suppliers) as flooring.

Adjacent to the living room, the author's indoor/outdoor room is an ideal place to sit and enjoy conversation and watch the plants grow. (Photo by author).

The design for this space uses a butterfly roof; note the sunken pit at front and the use of brick flooring. (Photo by author).

Using the earth (the prime natural material) is a good idea too; sink the building in the ground to have the earth act as an insulator. But no matter what materials you use, remember to consider climate. In temperate areas use heavyweight materials on east and west walls to restrict the flow of solar heat during summer days and thus provide coolness. Use lightweight materials for north and south walls so there is good cooling at night; the south wall should be glass. In arid regions use a heavyweight north wall to give cool conditions during hot days for daytime living and to retain heat for the night. The south wall should be of lightweight construction and glass to allow rapid heat buildup by day but coolness at night. Some shade from daytime glare will be necessary on the south wall. (See Chapter 8).

4 Building the Indoor/Outdoor Room

WHETHER YOU DESIGN AND BUILD YOUR OWN ROOM OR HAVE someone (designer or contractor) do it for you, some basic information about construction and materials will help you. It will also provide a background for you so you can construct a functional room, for many times the design of the room depends upon stock-sized materials like lumber and glass if you want to build cheaply (and who doesn't!).

ATRIUM

TERRACE

MAIN HOUSE

PERGOLA

SCULPTURE

POND

SUMMERHOUSE

DECK

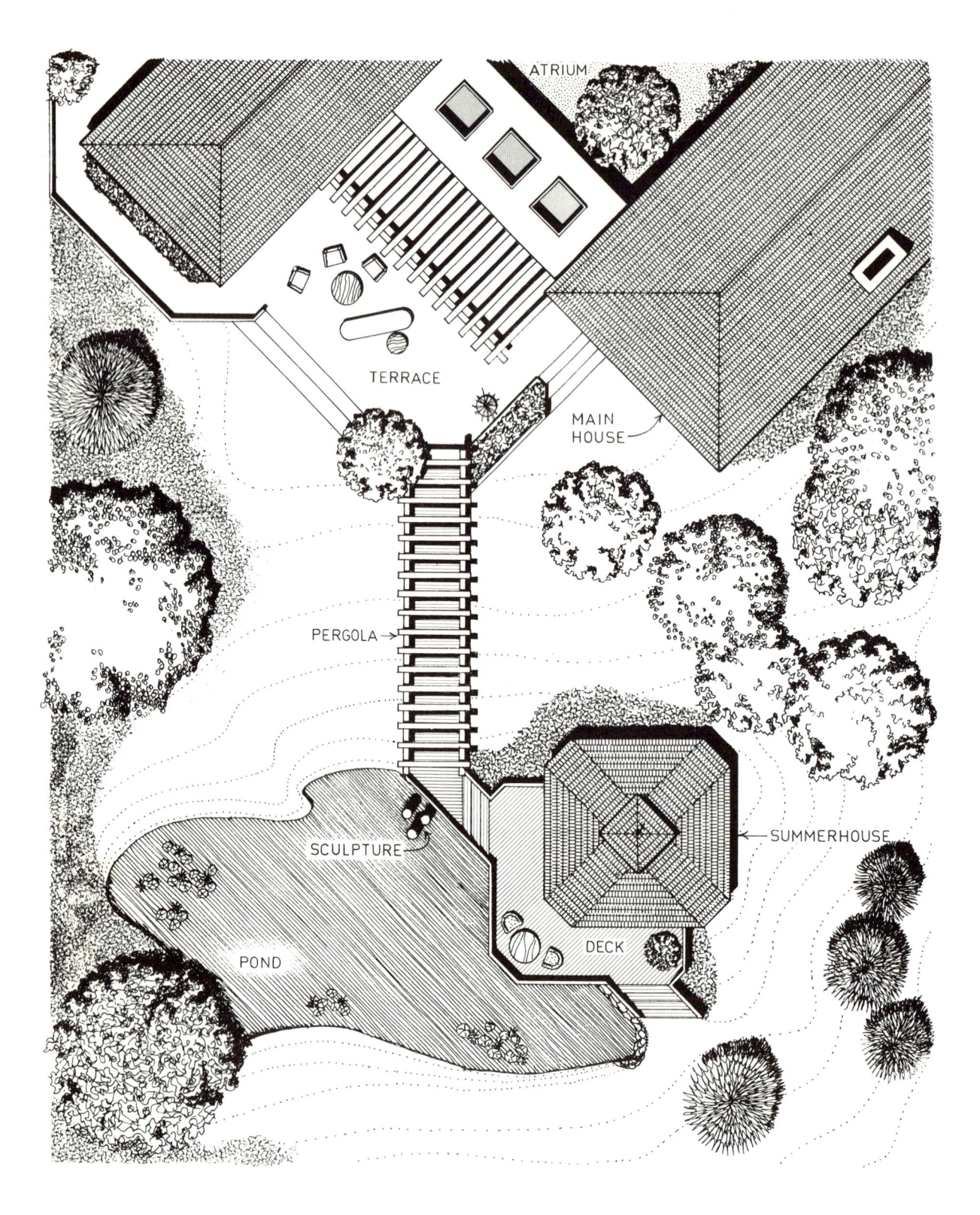

Site Plan

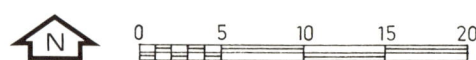

N   0 5 10 15 20

Have patience in selecting the final design. As mentioned in the previous chapter, make sketches; this is helpful, and it is easy to correct errors on paper but costly if you decide to change once building is under progress. Because we are trying to utilize the sun and nature to the utmost, certain factors and ways of doing this must be included in the plans before you start.

You will also need to know something about footings and foundations. And be sure you check local building codes before you start; there may be restrictions as to kind of foundations, drainage, and so on. In most cases expensive electrical work or plumbing will not be necessary. Battery-powered lighting can be used if necessary, and auxiliary space heaters can be used on very cold nights, but generally the room is a daytime rather than an evening area.

The following information is rudimentary and not to be considered comprehensive. It will provide some basic knowledge of sun-heated room construction and materials so you will know what to ask for when talking to carpenters or how to do it if you build your own room.

Footings

Footings and foundations anchor the building in the ground and are your first consideration (a footing is part of the foundation). They are necessary for any type of room and several types of footings are shown on page 42. Footings vary and may be of slab construction, or a footing-and-foundation wall, or a footing with a masonry wall. Whichever design you follow excavation of existing soil will be necessary. (Footings may also be concrete tubes or precast piers (sometimes called footings too).

So although designs may vary depending upon where you live and building codes, the following general plan can be used:

1. Drive twelve stakes 4 to 8 feet from the prospective dividing corners of the desired design. Then, using string, lay out the exact plan of the building from stake to stake.

2. Dig a trench around the desired perimeter of the proposed site. Make the trench approximately 2 feet wide and a minimum of 1 foot deep (or whatever building codes advise).

3. Once you have decided on the height of the foundation, use a level to be certain all stakes are on the same level. (*Note:* If possible, leave 3-inch holes in the base of the foundation about every 6 feet so water can run off to a lower grade. The drainpipe should extend all around the exterior of the room at the base of the footing in the trench.)

4. Foundation framing equipment is usually available on a rental basis. If not, use 3/4-inch plywood to frame the foundation. The width of the footing should be 8 inches (or whatever local building codes require).

5. Reinforce footings with steel rods laid horizontally and vertically within the footing. Pound the vertical rods in the ground between the foundation framing and then, using wires, tie the horizontal rods to the vertical ones.

6. Leave 1/4- to 1/2-inch "D" anchor bolts (available at lumberyards) protruding from the top of the footing so there will be a base to secure the bearing plate. The plate should be laid approximately 1 inch inside the outside line of the footing.

7. Apply a vapor seal to the top of the footing to stop capillary action.

8. Be sure there is adequate drainage system at the base of and through the foundation or under the slab to carry off excess water.

9. For outside drainage, place drain tiles at the base and through the footings, on 4 to 6 inches of rough gravel.

10. For inside drainage, plan a floor drain (optional). Before the floor is installed, locate the drainage heads in a low area of the flooring.

Concrete Floors

Concrete is an economical and durable floor because it resists water spill if there are plants and it retains heat. Always install a gravel base; this provides a solid level place for concrete to rest and helps to eliminate cracks in the concrete. Over the gravel install a plastic sheet which will act as a vapor barrier.

The floor is poured along with the foundation and should be at least 4 inches thick and reinforced with steel rods and wire mesh. Again, be sure the ground is absolutely level; otherwise concrete can eventually develop cracks.

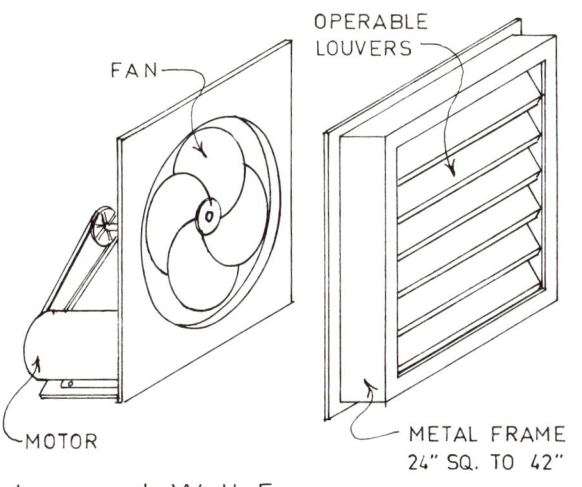

Ventilators

Louvered Wall Fan — FAN, OPERABLE LOUVERS, MOTOR, METAL FRAME 24" SQ. TO 42"

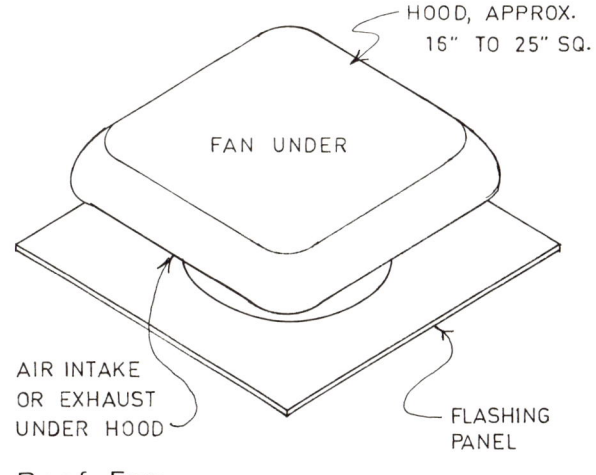

Roof Fan — HOOD, APPROX. 16" TO 25" SQ., FAN UNDER, AIR INTAKE OR EXHAUST UNDER HOOD, FLASHING PANEL

Skylight w/Fan — TRANSLUCENT PLASTIC DOME, FAN & MOTOR, CRANK OPERATED LID, APPROX. 14" SQ. OPENING

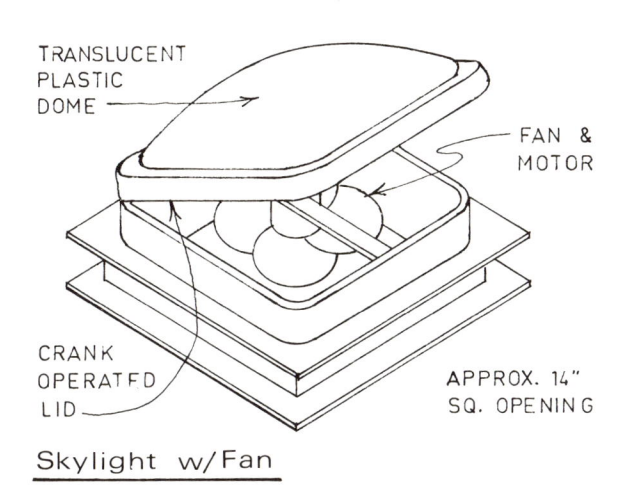

Wall Fan — 10" OR 12" SQ. GRILL, DIRECT OR PIPED VENT, PULL CHAIN, SHUTTER, FAN & MOTOR

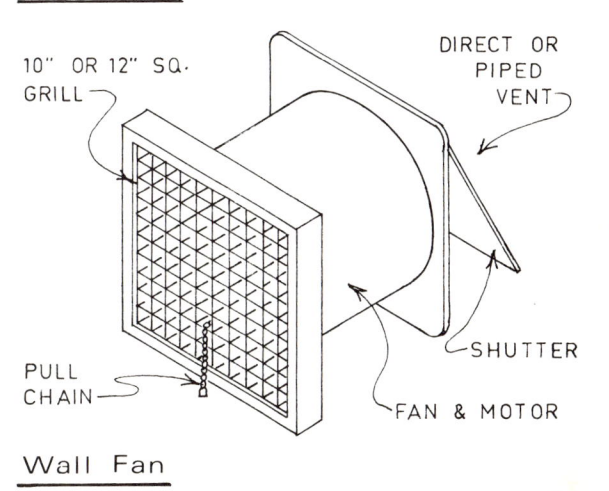

Drainage

Solid Floor — DRAIN COVER, FLOOR SHOULD SLOPE SLIGHTLY TOWARDS DRAIN, CONCRETE, GRAVEL, DRAIN PIPE

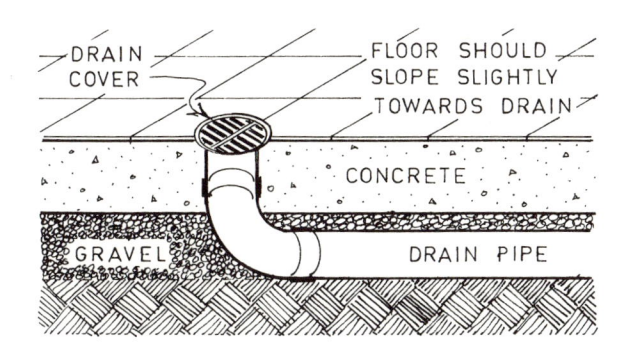

Gravel Floor — GROUND IS SLOPED TOWARDS CLAY PIPE IN TRENCH, GRAVEL, 3" OR 4" DIAMETER CLAY DRAIN PIPE

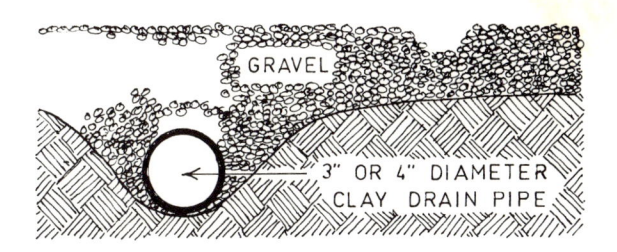

Footing w/Masonry Wall

1. DIG TRENCH & MOISTEN SOIL TO FIRM IT
2. POUR 1/2 OF FOOTING THICKNESS & LAY 1/2" STEEL REINFORCING BAR (REBAR)
3. QUICKLY POUR 2ND HALF OF CONCRETE & ALIGN VERTICAL REBAR
4. WHEN FOOTING IS SET CONSTRUCT WALL & FILL CAVITIES W/CONCRETE

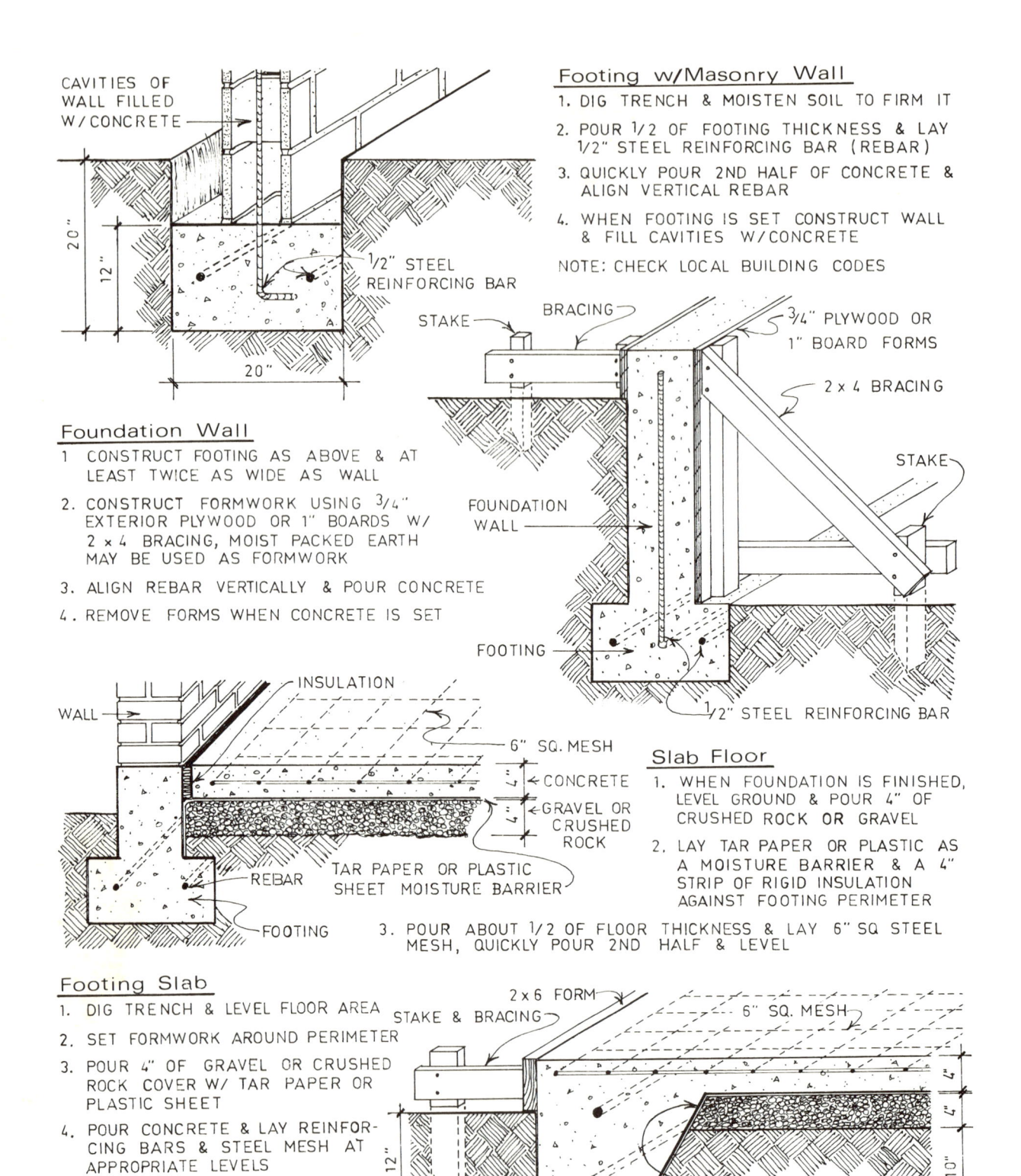

NOTE: CHECK LOCAL BUILDING CODES

Foundation Wall

1. CONSTRUCT FOOTING AS ABOVE & AT LEAST TWICE AS WIDE AS WALL
2. CONSTRUCT FORMWORK USING 3/4" EXTERIOR PLYWOOD OR 1" BOARDS W/ 2 x 4 BRACING, MOIST PACKED EARTH MAY BE USED AS FORMWORK
3. ALIGN REBAR VERTICALLY & POUR CONCRETE
4. REMOVE FORMS WHEN CONCRETE IS SET

Slab Floor

1. WHEN FOUNDATION IS FINISHED, LEVEL GROUND & POUR 4" OF CRUSHED ROCK OR GRAVEL
2. LAY TAR PAPER OR PLASTIC AS A MOISTURE BARRIER & A 4" STRIP OF RIGID INSULATION AGAINST FOOTING PERIMETER
3. POUR ABOUT 1/2 OF FLOOR THICKNESS & LAY 6" SQ STEEL MESH, QUICKLY POUR 2ND HALF & LEVEL

Footing Slab

1. DIG TRENCH & LEVEL FLOOR AREA
2. SET FORMWORK AROUND PERIMETER
3. POUR 4" OF GRAVEL OR CRUSHED ROCK COVER W/ TAR PAPER OR PLASTIC SHEET
4. POUR CONCRETE & LAY REINFORCING BARS & STEEL MESH AT APPROPRIATE LEVELS
5. LEVEL FLOOR & REMOVE FORMS WHEN CONCRETE IS SET

Concrete Details

To build a concrete floor you will need wooden forms, that is forms to hold the concrete until it sets. You can build your own forms, but in most parts of the country you can rent them, or better yet, use steel stake forms, which can also be rented. The forms must be put in place absolutely level, with the top wood board floor-level. Reinforcing steel rods must go around the footing; these support the weight of the building and anchor it.

You can mix your own concrete for a small area (5 x 10 feet) by renting a power mixer and putting in cement, sand, and gravel. The usual proportion is 1 part cement to 2 parts sand to 2 parts gravel or aggregate. Into the mixer put water, then gravel, then sand, finishing with the cement. The job must be done quickly, before drying sets in, and the floor must be smoothed at one time.

For larger areas buy ready-mixed concrete and have it delivered to the site. A truck generally holds about 7 yards; the concrete runs from the chute directly into the site where you have the forms in place. (You can also have companies pump concrete up a hill or into inaccessible places, but that will of course cost more money.) Ready-mixed concrete is a boon and convenient, but you must be prepared to work fast, within a 30-minute period, or you will get charged overtime. Try to have three men on hand; one to guide the chute (or the truck driver might do this), and two men with floats and trowels to get the mix in place. If concrete starts to set before you have finished the pour, you are in trouble, so keep working, and fast. As the pour is being done, take sticks and poke into the footings to be sure concrete gets to all voids in trenches.

Distributing the concrete at a uniform level in the form area is known as screeding. This is done with a screed board (two pieces of wood nailed to each other—a 2 x 4 with a 1 x 2 handle). Use the screed board (and wear rubber boots) to level the concrete as it is poured: push and pull the board to achieve the level surface. Keep a flat-nosed shovel on hand if the pouring gets ahead of you and too much concrete is dumped at one time. Next, tamp down the concrete (this step can be omitted); run an expanded metal screen over the concrete to level the slab and bring water, sand, and cement to the surface. Use a float (a 2 x 6 wooden board with a handle) to further level the concrete. Work it over large areas while the concrete is still wet. Do not dig it in; use a light touch. Be sure the handle of the float is long enough to reach the middle of the area of the slab from the outer edge. Use the float again when the concrete is sugary or somewhat set, and work in wide sweeps to level the slab.

(*Above*) The support for this indoor/outdoor room are concrete columns; four columns support the weight of the concrete aggregate floor. (Photo by Richard Dilday).

(*Left*) Here, the concrete floor is shown (at bottom) with one of the column supports. (Photo by author).

Steel troweling is the final step; this seals and waterproofs the slab and gets rid of minor defects. When the concrete is set (time depends on climate and type of pour), remove the wooden forms. For a few days, especially if the weather is hot, cover the area with plastic or burlap, keep the concrete sprinkled with water so it cures slowly to give you a strong floor.

The completed indoor/outdoor room; this adjoins the main house. (Photo by author).

Brick Floors

While concrete is an excellent flooring material, brick too has its merit. It is always handsome, relatively easy to install and weathers beautifully with time. It resists stain and can be installed in many designs to add accent to the indoor/outdoor room. In cost, it is somewhat more expensive than concrete.

If you are using brick the room needs substantial foundation as outlined previously. While brick may seem difficult to lay, it is actually a question of having patience rather than experience.

There are many kinds of brick for flooring! Common brick is generally available with pit marks on the surfaces and sand mold brick is smooth textured and slightly larger on one face than the other. If possible, select hard burned rather than green brick. It is generally dark red in color, not salmon which indicates an underburned process and less durability.

Bricks are set in mortar on a concrete base and the process of installation is the same as for building a wall. Use a thin mortar for laying the bricks and a heavier cement and grout between the bricks. Many people avoid doing their own brick work because they believe cutting brick is difficult. It is, in fact, quite easy. Use a cold chisel or a brick hammer for making irregular cuts and for trimming. Cut a groove along one side of the brick with the chisel or hammer and then give it a final severing blow. Cut the brick on a solid level surface such as a piece of wood. Smooth uneven bricks edges by rubbing with another brick. Common brick should be damp but not wet when laid.

Walls

Wall construction generally involves a combination of glass along with wood, aluminum, or masonry as its supporting element. South walls will require more glass than the other walls. Indeed, it is best to use as little glass as possible in east and west walls. North walls are generally the fourth house wall. Just how large your windows should be depends on the size of the walls. In many cases, long vertical spans of glass to admit light are quite handsome in combination with masonry. Stock-size aluminum or wooden windows can also be used but the total effect is less handsome.

Often, you will be working with hollow core concrete block. These are lightweight and give good insulation, more effectively deaden sound, and are easy to lift. The most common size is 16 inches long, 8 inches high, and 8 inches wide. Foot-square and 4-inch blocks are also available. In addition to the standard blocks there are half, corner, double corner, bull-nose, and channel block to help you build easily.

For a textured look, try split block or slump block. The split form has a rough face and slump block has good dimensional quality. You can also use two types of blocks in one wall to create an interesting pattern; for example, alternate a row of 8-inch blocks with 4-inch ones.

There are different ways of treating the mortared joints. A tooled joint is sort of a half round cove or squeezed joint, where mortar is allowed to show between the joints. A raked joint produces a sharp relief; do this by cleaning the mortar from the joint to a depth of ½ inch or less.

Laying Block

A block wall needs a substantial footing of concrete. The foundation may be 18 or 24 inches (check local frost lines). Pour the foundation in forms; when it is completely dry, start the wall work. Use a mortar mix: 2 shovelfuls of masonry cement to about 5 shovelfuls of mortar and sand. Use just enough water to make the mix plastic so that it clings to the trowel and block without running or squeezing down when you lay the block. As you work you will learn the right consistency for the mortar. Lay out the blocks on the foundation without mortar, and shift them around until they fit. The idea is to save you having to cut blocks. Keep spaces between the blocks no wider than ½ inch, no narrower than ¼ inch. Clean the foundation and wet it down. Now mix mortar, or use a plastic cement or a premixed mortar (you add only water). Now trowel on a 2-inch bed of mortar, and seat the first block. Tap it into place with the trowel handle. Repeat the process, putting mortar on the inside end of each succeeding block.

For a sturdier wall, lay the block on a footing that is still in a plastic state (i.e., has the consistency of mortar). This first course of blocks is then solidly attached to the foundation. When the concrete foundation has become like the consistency of mortar it is ready for blocks, but first, as in any foundation, position blocks on the ground along the side foundation so you have enough blocks and little cutting is necessary. Then seat the block about 2 inches deep into the mortar. You want the foundation concrete still to be pliable so you can level blocks. Start at the corner with a level and square-shaped corner block, and trowel mortar in strips on outside edges of the first course. Do one block at a time and tap it into position. Always be sure it is level and flush with the block beneath. Keep courses even with a mason line. (See brick section for mortar properties.) Put down just enough mortar for one block at a time.

If you place a wall where there is a drainage problem, you will have to put drain tiles along the outer edge. Slope the wall about 1 inch for each 15 feet.

For tall walls (over 5 feet), use reinforcing rods set vertically and solidly in the concrete foundation. Space them according to local building codes. Lay the first course of the wall in wet concrete, and then drive rods through the cores. The holes for rods must match the holes in blocks, so alignment is vital. Don't forget to leave framing space for window openings.

Brick

The beauty of brick cannot be denied; it is a natural material that harmonizes well with most outdoor situations. Further, brick stands the test of time and becomes beautiful with age. Brick is now offered in a multitude of patterns: thin, thick, colored, and in various shapes.

The average brick wall is 8 or more inches thick (two bricks wide) and requires steel reinforcing rods in mortared joints at frequent intervals. Very tall walls have to be reinforced about every 12 feet with a brick pier or pilaster. This type of construction requires the help of a professional mason. You can dictate the pattern to suit your tastes, but the actual building of the wall (unless you are very handy with tools) generally must be farmed out. However, for those who want to try constructing their own brick wall, you will need a pointed trowel for buttering mortar; a broad-bladed cold chisel; and a hammer, level, and carpenter's square.

Common brick must be damp to be laid. To hold the mortar you need a mortar board, which is a piece of wood, for example: ¼-inch plywood. Scoop the mortar (enough for only a few bricks) from the board with the trowel, and spread it over the top course of bricks. Put each brick in place, trim away mortar to butter the end of the next brick, and continue until more mortar is needed. Bricks should be set in perfect alignment; tap them into place gently. Build the ends of corners first in steps because this will make it easier to set the next bricks in line. Be sure to use a strong guide line, that is, a nylon line, to guide you in laying the bricks. Anchor the ends of the line into the mortar joints. Before the mortar sets, trim away loose bits and smooth off all joints.

Mortar for brick laying is a mixture of cement, fine sand, and water, some lime added for plasticity: 2 parts Portland cement, 1 part fireclay or lime, and 9 parts garden sand. Supplies are at hardware and lumber stores. Do not use beach sand which contains salt and will resist hardening.

Concrete

We have mentioned the advantages of concrete for flooring, and for walls concrete is again a good choice. It is extremely strong and is easy to clean and again is a natural heat retaining material. Its one disadvantage is that you must have wooden or steel forms (from rental companies) to do the job. These precise forms are necessary and careful pouring of the concrete is essential for an attractive window wall.

Aligning and building the forms for a concrete wall takes more time than the concrete pour so it is best to rent the forms. Be sure they are properly braced with 2 x 4's so they resist the pressure of wet concrete without collapsing. Include reinforcing rods along with steel mesh to add strength. Building a tall wall for your indoor/outdoor room requires staging and ramps to permit the loaded wheelbarrow to be rolled to the top of the forms for dumping. This is backbreaking work so try if you can to use ready mix concrete delivered by truck. Then a chute is placed into the forms and pouring is done quickly. Pour the concrete in place 6 to 8 inches in a continuous pour and then tamp in place with a shovel. It will be necessary to have another person on hand to assist in the pouring operation because time is of the essence once the ready mix truck arrives. Be sure to remember to form-up for window openings. Curing the concrete and removing the forms follows the same procedure as for floor work.

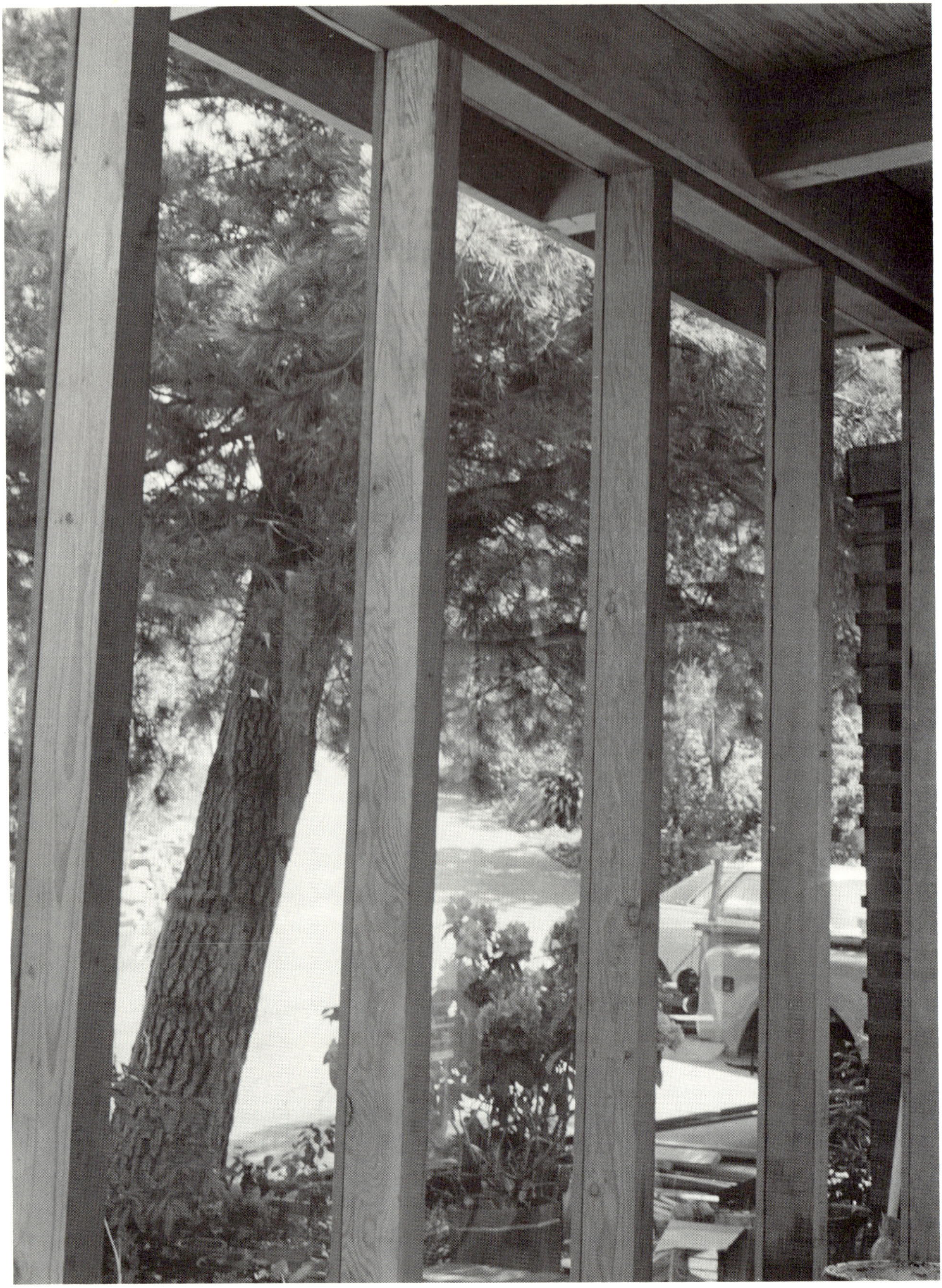

Glass and Glazing

Because the indoor/outdoor room is built with some glass, it is wise to know something about the material. Thermopane glass is highly recommended, to conserve as much heat as possible in winter. This is two pieces of glass hermetically sealed, with an air space in between. These units come in clear glass in thicknesses of ⅛ inch, ¼ inch, and so on or in patterned glass in different thicknesses. Thermopane is available in standard sizes ranging from 16 x 20 to very large sizes, and before you start to design or build it is wise to have a list of the standard thermopane sizes. Custom units made to size are 25 to 40 percent higher and take much more time to get than standard stock sizes.

Glass is available as ⅛ inch or 3/16 crystal quality. This is a rolled glass, which means it will have some waves, but this is not objectionable in small sizes. In large sizes it is better to use ¼-inch polished plate glass. This glass is polished on both sides to prevent waviness and presents a clear image when you look through it.

All glass and glazing should be done by a professional because it is extremely important that there are no air leaks around the glass units. Drafts can drain away a great deal of accumulated daytime heat and leave a room cold by dusk.

When using wood frames for glass (as most often is the case) be sure the wooden members are rabbetted to accommodate the thickness of the glass. This should be one-piece construction and not two separate pieces of wood. Glass is set into the rabbett (generally ¼-inch deep), and then putty is applied around the edges and a wood molding is placed on top of the glass and putty and nailed to the frame. This type of glazing virtually assures a sealtight window.

When glazing glass in wood be sure the wooden member or framing is rabetted so glass lies flat in a one-piece member. Putty is then applied around edges of glass and wood strips nailed to the support members to sandwich glass in place. (Photo by author).

Roofing

The roof is one of the more important parts of indoor/outdoor rooms. Its design may be gabled, A-frame, vaulted, sawtooth or simply at a 45 degree angle (against a house wall). The roof should contain at least 30 percent glass to admit sufficient light and this may be in the form of skylights or clerestory windows or simply domes.

Glass in metal or iron frames is frequently used but it is rare to find a completely leakproof unit. Clerestory windows seem a better solution because they are set like ordinary windows and should not leak. Flat glass set in wooden panels is yet another way to approach roof design and these like peaked or canted skylights are difficult to build and, further, generally leak water. Plastic domes of many different shapes are available too and seem to be airtight and leakproof.

No matter what you decide to use, wood or metal framing to accommodate the glass or domes is necessary; this in turn must be adequately protected with metal flashing to eliminate the possibility of water seeping into the structure.

In any area where glass is in the ceiling, building codes require the use of tempered or wire glass. This is much more costly than standard window glass. In clerestory windows, double strength or thermopane window glass is generally allowed but to be sure check with local building offices.

Glass must be put in place properly and this is done with putty or equivalent glazing compounds. The glass sits on the putty and is the cushion for it. To further assure an airtight enclosure plastic cushioning tape is frequently used.

Tongue and groove lumber is usually the other component of roofing; these are interlocking boards to assure a solid roof as well as an airtight and leak-proof one. In addition shingles or tar-and-gravel are applied over the ceiling boards for absolute weather-proofing. My indoor/outdoor room has ¾-inch hemlock, a very durable strong wood, and has proved very satisfactory.

Roofing can be skylights with glass or as shown here, domes. The dome is simple to install and is generally leakproof as well as airtight to eliminate heat escaping. (Photo by author).

Domes

Domes are a product of American ingenuity; custom-made skylights with glass are expensive. Premolded acrylic domes for ceilings are generally moderate in price. For example, a 37 by 37 inch clear plastic dome costs about $75. While the clear dome is generally used, opaque or tinted domes are now available also. Because of its cost and because domes are more air proof and leak proof (as a rule) than glass skylights, it pays to investigate this facet of ceiling construction carefully. Many times, depending on the indoor/outdoor room, domes can be the answer for admitting light. Two or three moderate-sized domes would be all that would be needed in an average room to give sufficient light.

When people think of these units (which come in various stock sizes) they think of a dome shape but the word is used loosely because domes also come in peak shapes and angular shapes. Each style has a different character and attitude about it and what you use will ultimately depend on the design of the building itself.

Domes are easy to work with since they set into precut wooden frames and actual installation requires adequate metal flashing to eliminate any possibility of leakage. Again, remember that domes come in stock sizes; these are cheaper and more readily available than custom domes. So when designing your indoor/outdoor room have dome sizes on hand. They are available at most local glass dealers.

A conventional standard size (41 x 41 o.d.) plastic dome. Note: Domes also come in peaked designs. (Photo by author).

5 Making the Indoor/Outdoor Room Comfortable

A SUN-HEATED ROOM IS A FUNCTIONAL ROOM, SO INCLUDE FURNIture. Benches, tables, and seating groups of many designs are available from suppliers. Or of course you can build your own; there is intrinsic beauty in custom-made furniture. Also, when furnishing the additional space include some plant furniture: plant stands, benches, and so forth, and some plants to decorate the area and provide the necessary green accent.

There are so many different kinds of furniture for casual living that it is impossible to discuss them all here. The best advice I can give is to shop thoroughly and pick and choose carefully so you have a total room environment. Do not just take anything; get commercial furniture made of wicker or bamboo, or make your own.

Benches, Other Furniture

Generally, outdoor or patio furniture is large and too bulky for indoor/outdoor rooms. Also, this furniture, made of commercial plastics or other synthetic materials, is rarely handsome, in my opinion. Seek distinctive pieces of wicker or bamboo, which blend well into outdoor rooms. Or make your own built-in furniture. Benches of a beach-type design are being manufactured again and are excellent in looks and comfort. Use furniture that is simple and in scale with the size of the room. Proportion or balance is vitally important in creating a total pleasant place. Do not forget that furniture will be subject to extreme sun; wicker, bamboo, and redwood all hold up well under sun.

Commercially made wicker or bamboo furniture is fine and attractive in a room, but I prefer built-in benches or seating areas that become part of the total room and occupy less space. For example, a slatted bench built to fill a specific area can be stunning with brightly colored canvas throw pillows on it. If you are halfway handy with tools, you can build redwood benches or shelves for plants. Nothing elaborate is necessary; strive for simple clean lines in your do-it-yourself furniture. There is a galaxy of simple good-looking designs in garden furniture you can make.

Plant Furniture

You can make these units yourself or buy them. Suppliers have some lovely plant stands and pedestals, for example, attractive wrought iron plant

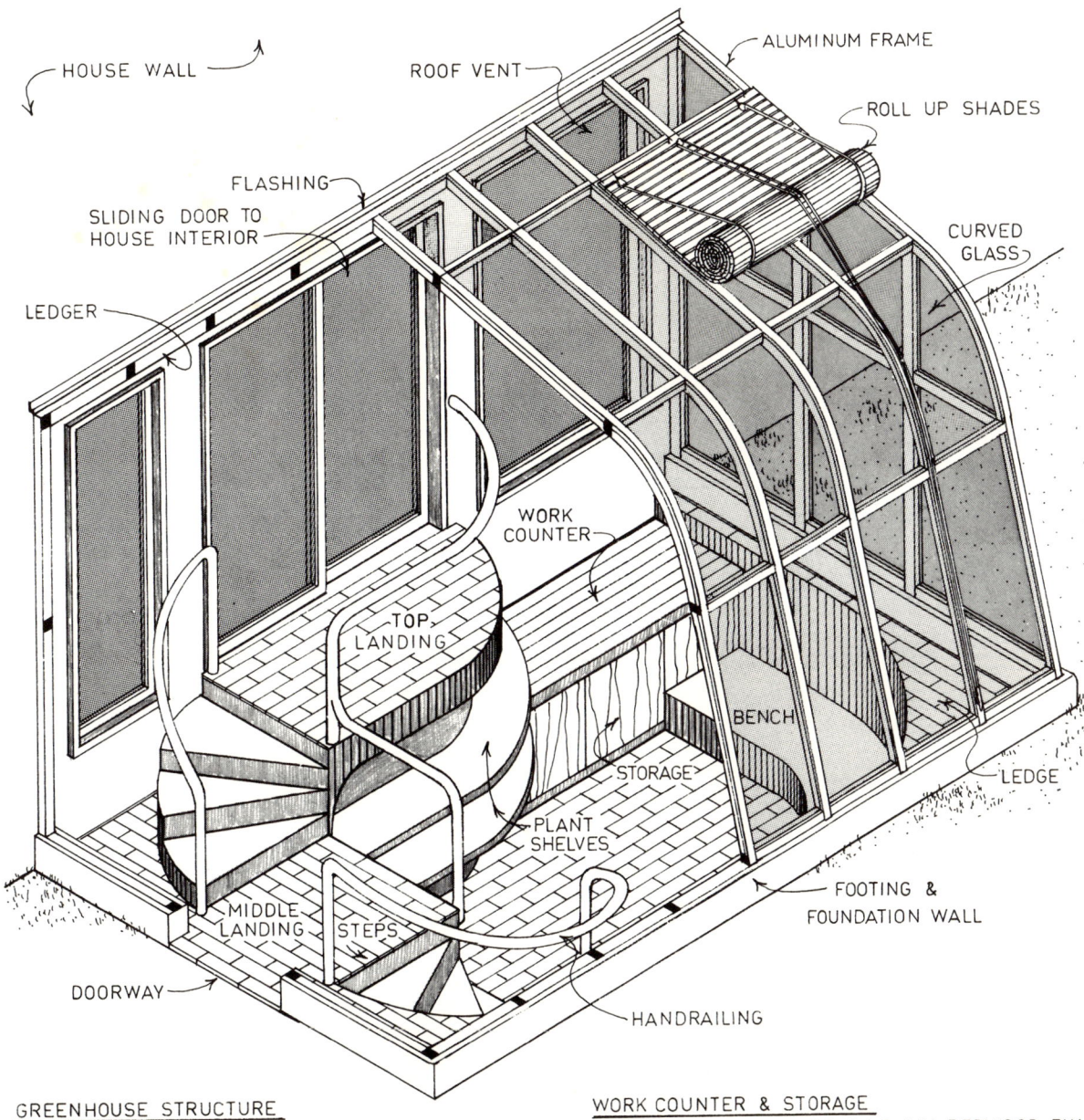

GREENHOUSE STRUCTURE
PRE-FABRICATED LEAN-TO, ALUMINUM FRAME & GLASS, CURVED EAVE & VENTS AT RIDGE 10'-6" WIDE, 17' LONG, 9' HIGH, SET AGAINST HOUSE OVER RECESSED FLOOR

FOUNDATION WALL, FLOOR, STEPS & BENCH
POURED-IN-PLACE REINFORCED CONCRETE, BRICK TILE FLOOR & LANDINGS, OTHER SURFACES TINTED & TEXTURED

HAND RAILINGS
CURVED STAINLESS STEEL PIPE, 2½' DIAMETER

WORK COUNTER & STORAGE
2 x 4 REDWOOD TOP, ROUGH-SAWN REDWOOD PLYWOOD CABINET ON CONCRETE BASE

PLANT SHELVES
LAMINATED REDWOOD ON STEEL SUPPORTS SET INTO CONCRETE WALL

DOOR
ALUMINUM FRAME W/GLASS JALOUSIE WINDOW

SUN SHADES
WOOD OR ALUMINUM SLAT ROLL-UP SHADES

Recessed Greenhouse

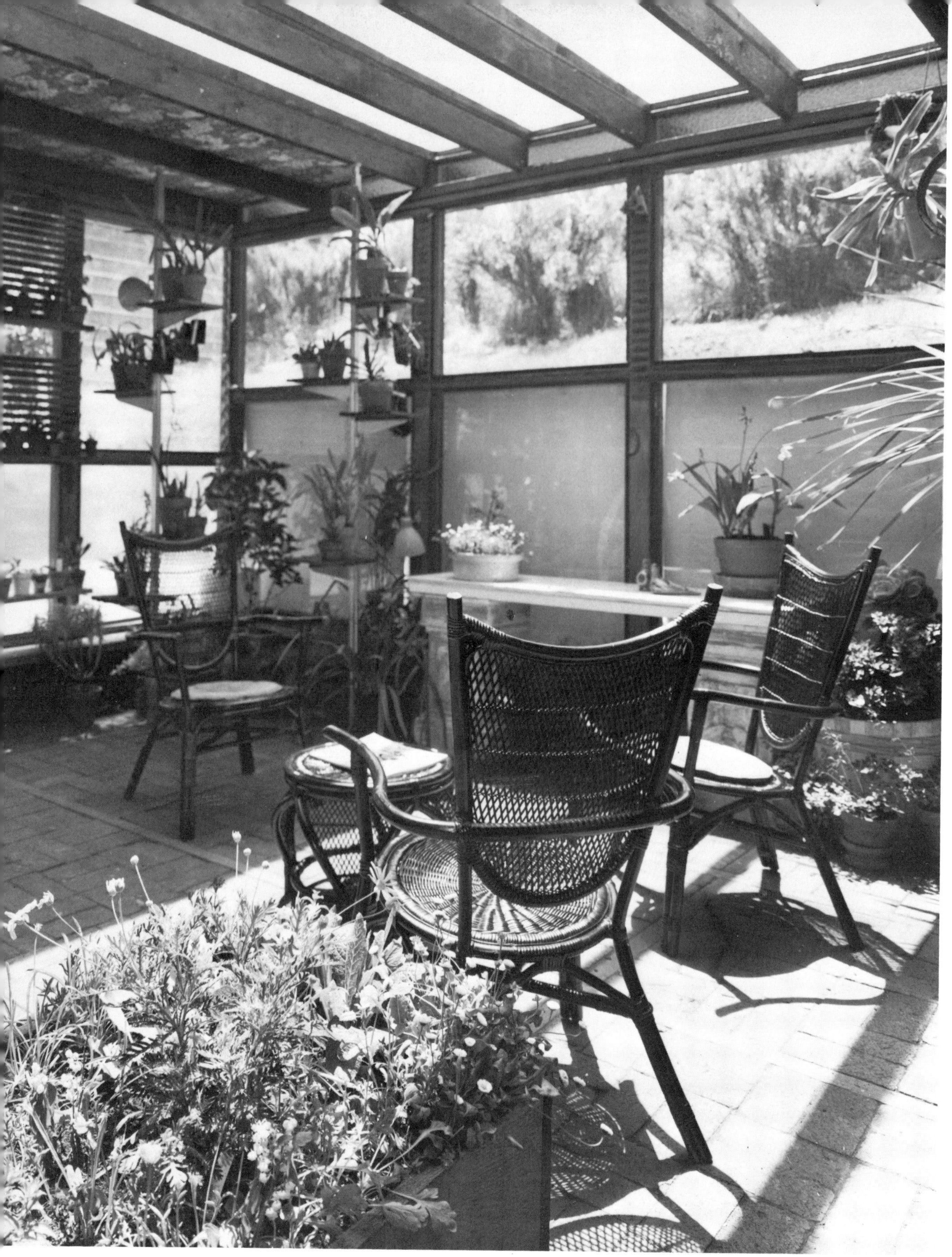

(*Above*) Wrought iron furniture is used for an outdoor gazebo. (Photo courtesy California Redwood Association).

(*Left*) Outdoor type furniture can be used in the indoor room; this is inexpensive wicker design. The seating area is quite pleasant surrounded by greenery. (Photo by Joyce R. Wilson).

stands that accommodate many plants. Plant pedestals and plant shelves can also be used to great advantage to display plants. Potted plants set on floors helter-skelter have little impact on the viewer, but plants on their own pedestals or platforms are dramatic.

Special containers such as jardinieres and decorative tubs also fit into the decor of the indoor/outdoor room. Planters and pots (at suppliers) range from terra cotta to gold-glazed tubs to ornate urns. Wood containers are suitable too, although some consideration should be given to design because many commercially made ones are ugly.

Besides room and plant furniture, containers and plants, there is still another vital part of furnishing, if your budget will allow it: a lovely small sculpture. Sculpture can add great beauty to the space, so indulge yourself.

Super Plants

In Part Two you will note that almost any kind of plant can be grown in a greenhouse. However, in the indoor/outdoor room, where plants are not the prime consideration, it is wise to choose special types of mature plants that will grow easily, add beauty, and always be pleasant to look at. Some plants are better than others in these roles; they need less care and make a better display than most plants, which is what this section is about.

Use a few healthy mature plants (called "specimen" in the trade) rather than many small pot plants, although it is fine to bring in small seasonal flowering plants. Look for treelike plants such as *Ficus benjamina, Clusia rosea, Schefflera actinophylla,* and one of my favorites, *Caryota mitis* (fish-tail palm), which seems to grow without any personal attention. Buy branching plants like *Dracaena marginata* or silhouette beauties such as large columnar cactus (Cereus). Avoid frilly, dainty plants that give a bushy effect; this is fine for outdoors where mass is needed but not indoors where more refined lines are necessary to complement the design of the room. Avoid small ferns because they require too much attention. Palms also can be difficult indoors, except for the bamboo palm (*Chamaedorea erumpens*) or fish-tail palm. Select the sculptural and dramatic for display in the indoor/outdoor room. Again, you do not want too many plants, but rather a few well-chosen ones. Care will be at a minimum and decorative accent at a maximum.

Large plants may cost a great deal of money, but they are easier to grow and require less care than small plants because they are at the peak of health and are strong and robust. They can, if necessary, take abuse and still survive beautifully for a few weeks. Feeding and watering on a bimonthly schedule will suit most specimen plants during spring and summer. In fall and winter just keep soil barely moist and feed plants only once each season. Too much feeding can burn plant foliage.

Space does not allow me to list here the many fine indoor plants. Shop at stores or refer to a good house plant book. Remember: select the biggest and best you can afford and plants with special characteristics such as a lovely branching habit or sculptural growth to make a dramatic statement in your indoor/outdoor room.

Large plants are best for indoor/outdoor rooms. Here ferns and palms (but not too many) complement the scene. (Photo by author).

Suitable Conditions for People and Plants

The indoor/outdoor room requires a comfortable temperature and humidity for both plants and people. Quite naturally in winter nights will be cool—say 50°F—and this is fine for most plants just mentioned. However, it might be too cool for most people but generally these rooms would not be used during the very coldest months (January and February, for example). If temperatures in your regions are very cold in winter, then some auxiliary artificial heating will be necessary.

In summer, the goal is to keep the indoor/outdoor room or garden room cool because excessive heat exhausts both people and plants. Masonry materials—and we have stressed these throughout—for construction help greatly and proper ventilation and shading facilities can temper the heat considerably.

In summary, the indoor/outdoor room if built properly should function as a self-sustaining area with little or no artificial heat required. But few people are perfect and there are bound to be some problems. Live in the space a while to determine if it is drafty or too cold in winter, too hot in summer. Once you know the faults it is easy to correct them. Hedges and screens can be built to thwart strong winter winds; overhangs (if not already in place) can be constructed to temper the sun in summer. Faulty glazing of glass which creates air leaks and heat loss can be remedied with new adhesives. In short, minor faults can be corrected without too much expense or panic to make the indoor/outdoor room a comfortable place most of the year.

Part Two: The Greenhouse

6 A Place for Plants

IN PART ONE WE DISCUSSED THE INDOOR/OUTDOOR ROOM; THIS section deals with structures for plants. Generally, when we think of a place for plants, a greenhouse comes to mind. Actually, there are several types of places for plants: (1) the pit greenhouse, (2) the lean-to greenhouse, (3) the geodesic dome, and (4) do-it-yourself buildings for plants (perhaps prefabricated, lean-to greenhouses modified). The greenhouse may be attached to the house, for example, the lean-to or a detached space such as the pit or dome. Each has advantages and disadvantages.

Essentially, each of these units may serve the same purpose, but each is different in design depending upon climate and use (see Chapter 8). However, for our purposes, all have some things in common: they use the sun's energy as heat, and they are built of heat-retaining materials to lessen the cost of artificial heat and apparatus. Each type of greenhouse is discussed in the next chapter.

A greenhouse is a place to work with and grow plants. It can be a display greenhouse or an experimental hobby growing area. You can grow cut flowers to get a head start on spring, have exotic orchids or other plants, or use the greenhouse for propagating plants. The insulated pit calls for plants that will tolerate cool nights if necessary. The lean-to greenhouse or the dome is fine for flowering plants. The difference is in the construction details and design. Each has different heat requirements too. The lean-to or the dome will need some artificial heat; a pit greenhouse does not. The design of your greenhouse is governed only by your imagination.

The Pit Greenhouse

North American Indians grew vegetables in this kind of structure, and in the *McIntosh Book of the Garden*, 1853, there are numerous sketches of pit greenhouses dating back hundreds of years. The principle of the pit greenhouse is sound and works well but through the years has been overlooked. Yet, essentially it is a nature-working greenhouse, where climate works for rather than against you.

The pit structure is partially sunken into the ground at about 4 feet; one side is glass (south) pitched at a 45-degree angle, and the retaining walls are made of concrete or concrete block with the wall adjacent to the south wall painted a dark color. Three elements are involved here: the south *glass wall* or roof receives the full force of the sun, the adjacent *dark-colored* wall helps absorb heat for a long time, and the *concrete* below soil-line walls hold heat generated during the day. The floor is generally of earth or concrete, which allows warmth from the ground to be utilized. Because of these factors, enough heat is locked in for the night to keep temperatures well above freezing, even though it may be below zero outside. In very severe winter weather shutters can be used on the glass roof at night. If temperatures get too low, always remember

The pit greenhouse can be a lean-to or an A-frame such as this. Excavation is the difficult task for the concrete block walls extend down four feet. Properly insulated in winter with shutters this is a self-sustaining place for plants where no artificial heat is needed. (Photo by Matthew Barr).

Glass used at an angle gets the most benefit from sun's rays. The glass can be shuttered in winter or protected with insulated materials. (Photo by Matthew Barr).

that a small space heater (I bought mine at Walgreens for $8) will supply sufficient heat to keep the pit about 40°F on the coldest days. (See Chapter 7 for construction details.)

All in all, the pit greenhouse is a very simple and ingenious way of having plants in a space without the aid of any artificial heat. And there are more benefits. Because there is no artificial heat, generally, pests are at a minimum; there is no worry about overheating, which can kill many plants; and elaborate misting equipment is not necessary. Watering can be done in a half-hour when all plants are in one place.) If culture is good and plants are watered enough but not too much, the plants in a pit greenhouse are growing in almost perfect simulated natural conditions.

There are two types of pit greenhouses: one is double-eaved, with shelves on each side and the aisle in the center. The other has the entire lower part of the pit excavated below the soil line, with only one south glass wall, the ceiling. Technically the pit is detached from the house, but it is far better to use a house wall where possible to get the same heat and benefits as the house. With an entrance from a furnace room, the pit becomes an even warmer structure: warm air from the furnace radiates to the pit and keeps the temperature 5 or 10 degrees warmer when the furnace is on.

Just how you use your pit greenhouse depends upon your personal tastes. Not all or very many plants need humid, tropical heat 24 hours a day. Most plants, for example, cacti and succulents (fine house plants), rest in winter and need a lower temperature of, say, 45 to 50°F. You can use the pit greenhouse for tender woody plants that cannot take freezing outdoors. In spring you can start seeds and nurture seedlings until they are ready for outdoor gardens, or start perennials or store bulbs over winter (most bulbs need cold storage). The pit can also be used as a winter greenery by retarding autumn-blooming flowers and forcing spring-blooming kinds. Alpine plants, begonias, and cool-living plants such as camellias will flourish in the greenhouse.

House plants that like coolness, including philodendrons, azaleas, and some orchids, can grow all year in the pit. Herbs and vegetables will do fine too, although you should concentrate on cool-thriving herbs like rosemary, chives, or vegetables such as cauliflower and broccoli. In spring, lettuce, radishes, and other vegetables can be started and grown to perfection as warmth engulfs the pit in summer.

The Greenhouse (lean-to)

In America, the greenhouse is synonymous with the prefabricated-type design so often seen. These are either detached or attached. The detached greenhouse is disappearing fast from American properties. The attached unit is now more frequently called a lean-to because it uses one wall of the house in its design. Some units have curved glass, others use straight panes.

A prefabricated lean-to greenhouse with masonry walls foundation. Artificial heat is needed because of all-glass construction. Yet, cost is not exorbitant and the lean-to offers a fine place for plants. (Photo by Everlite).

The prefabricated greenhouse is made of aluminum channels, or redwood members and glass, and is delivered as a knocked-down (KD) package you assemble yourself. It still requires a concrete slab or footings. So does a do-it-yourself room. The redwood unit is now offered in several designs: octagonal, arched, hexagonal, and so forth.

The design of the prefabricated greenhouse may differ as mentioned but the basic structural components have remained the same: skeleton framing and glass. Generally this makes the structure costly to heat and drafty. Still, a commercial greenhouse can be used economically by modifying it so it is more functional—and less costly to heat. Or you can start from the bottom up and design your own greenhouse lean-to, utilizing principles outlined in this book. (Construction facts and drawings are in the next chapter.)

The Geodesic Dome

Geodesic dome greenhouses are not new. The Climatron in St. Louis and the three domes that make up the Mitchell Conservatory in Milwaukee, Wisconsin, are more than ten years old. If you have always admired domes, design your greenery on the dome principle because it is a good one. The hemispherical design best captures the sun's rays, as evidenced in old writings: quoting from a *Book of Garden Structures*, published in 1831: "The hemispherical figure is believed to be the kind of glass roof best suited to admit the sun's rays and ultimate in regard to the principles and perfection of form."

However, domes do have drawbacks. Mainly, the problem is in building an airtight structure. If made with all glass it will be excessively hot in summer, even with adequate ventilation. On the other hand, domes are inexpensive to build, excellent for small properties, and offer ample ceiling height for growing tall plants.

As with lean-to greenhouses, domes are now available prefabricated ready for building. These can be modified for your use, or you can build your own dome if you are handy with tools.

This handsome dome greenhouse is beautifully designed and a fine place for plants. It is situated taking advantage of the terrain. Overhangs protect the greenhouse from direct summer sun. Heavy tempered glass is used in window walls to further keep in as much sun's heat as possible. (Photo courtesy Roper IBG).

This is an inexpensive geodesic greenhouse. To use in severe climates a suitable foundation would have to be installed. Glass should be double paned to prevent heat loss. A combination of glass and plywood would be feasible too to make the area more self-sustaining. (Photo by Redwood Domes).

Placement and Size

Where you put the greenhouse and how large it is, whether it is attached or detached, is governed by the site; each case is individual. In any case, if you want to save heating bills, have the structure face south, where it can absorb as much of the sun's heat as possible (see Chapter 7).

At first, size may not seem important, but when you can build a greenhouse, say, 10 x 15 feet, for the same cost as an 8 x 10 structure, why not get the most for the least? Size is also important because sunheated buildings are best in small sizes. A large building is difficult to heat when you are using only the sun and an auxiliary heater. Height deserves consideration too because the south glass wall must be at a 20- or 40-degree angle to capture a lot of sun. You double your space with a tall structure.

No matter what kind of greenhouse you build, it should be an attractive adjunct to the home and garden and not an eyesore. Study greenhouse catalogs and especially photos of old greenhouses to determine the special design that best suits your needs and property. You will also need to know something about glass and plastics, glazing, and building to accomplish the ultimate personal greenhouse. And if you are not handy with tools, do not be scared—you can design it and have a carpenter build it. (Remember that even with a prefab greenhouse you must put it together.)

Greenhouses are generally part of the house. Thus, if properly designed, the place for plants can be a handsome addition to the house and become an extension of living space filled with lovely green plants. Some people enjoy the beauty of a greenhouse next to the kitchen, others like the idea of a greenhouse that extends a living or dining room, and more people are realizing the benefits of a small glass house next to the bedroom or bath. The bathroom especially is a fine place for greenery; if you want a tropical atmosphere, the bath greenhouse is ideal.

No matter which room the greenhouse adjoins, access to it is necessary. Sliding doors are inexpensive, work well, and can be opened during the day and closed at night. Metal doors look institutional, so, if possible, consider wood-framed sliding doors (more expensive but worth the extra cost). Also consider thermopane rather than single plate-glass doors. And, if you want some real charm, use French or casement-type doors. When you make your own greenhouse you can do all these things and have fun in the doing. One important consideration: many plants are thwarted from initiating bloom—poinsettias and carnations come to mind—if they receive any artificial light during the night; even 5 minutes can throw off the bloom cycle. So if the greenhouse is next to a living room, light from that room will affect your plants. Have some type of curtain or heavy drapery to draw at night if you want bloom on your plants.

The detached greenhouse is seldom seen today; they are too large, too expensive to heat, and more for commercial growers. Even pit greenhouses, where possible, are built adjoining the house, using one wall as part of the pit. The dome structure rarely is attached because its radical design makes it alien to most houses; however, a loggia or covered gallery from a room leading to a dome works well. This adds further dimension to the greenery and lends a note of sophistication to the total room plan. The gallery or loggia can be unheated, where cool-loving plants can be grown. This arrangement also opens up the possibility of hexagonal or geodesic structures as an extended part of the home.

How to Get Started

Because of the many ways of incorporating a greenhouse into the home, it is wise to do some thinking, planning, and rudimentary sketching before you start. Consider the total plan (design, materials, cost, and maintenance). See how it looks on paper. If cost is a factor, plan on a greenhouse with one wall of the house as its fourth wall. If you can afford the extra money, the gallery-attached unit is ideal.

To start the greenhouse idea in your mind, list what you want the greenhouse to do for you and your home:

1. Is it to be part of the house?
2. What do you want to grow in it?
3. What is the best size?
4. Do you need a view of green plants from a particular room?
5. Do you want it a very private place?
6. How much artificial heat can you afford for it?

When you have decided upon these things, jot down the choice and type (old or new) of materials. The shape may dictate the design, or often the design can dictate the materials. In any event, when you design your greenhouse, pit, dome, or whatever, do it with standard-sized materials in mind. Glass, lumber (on the even inch), and plexiglass all come in standard sizes. Nonstandard units cost 25 to 40 percent more, and delivery is months instead of weeks. If the greenhouse is going to have skylight construction, think about standard-sized plastic domes and industrial-type skylights (cheaper than custom-made), available in many designs as explained in Chapter 4).

7 Construction Know-How

EACH TYPE OF GREENHOUSE—PIT, LEAN-TO, OR DOME— MAY SERVE the same purpose, but each requires a different design and different construction know-how. The pit greenhouse may be the most difficult to do because of the excavation work; the lean-to is somewhat more simplified. (This greenhouse may be a total do-it-yourself project, or a prefabricated unit modified can be used.) The dome greenhouse requires exact planning and design, because the actual building may be more complex than the other types of greenhouses.

Construction differs, but all these units have things in common, the

Section

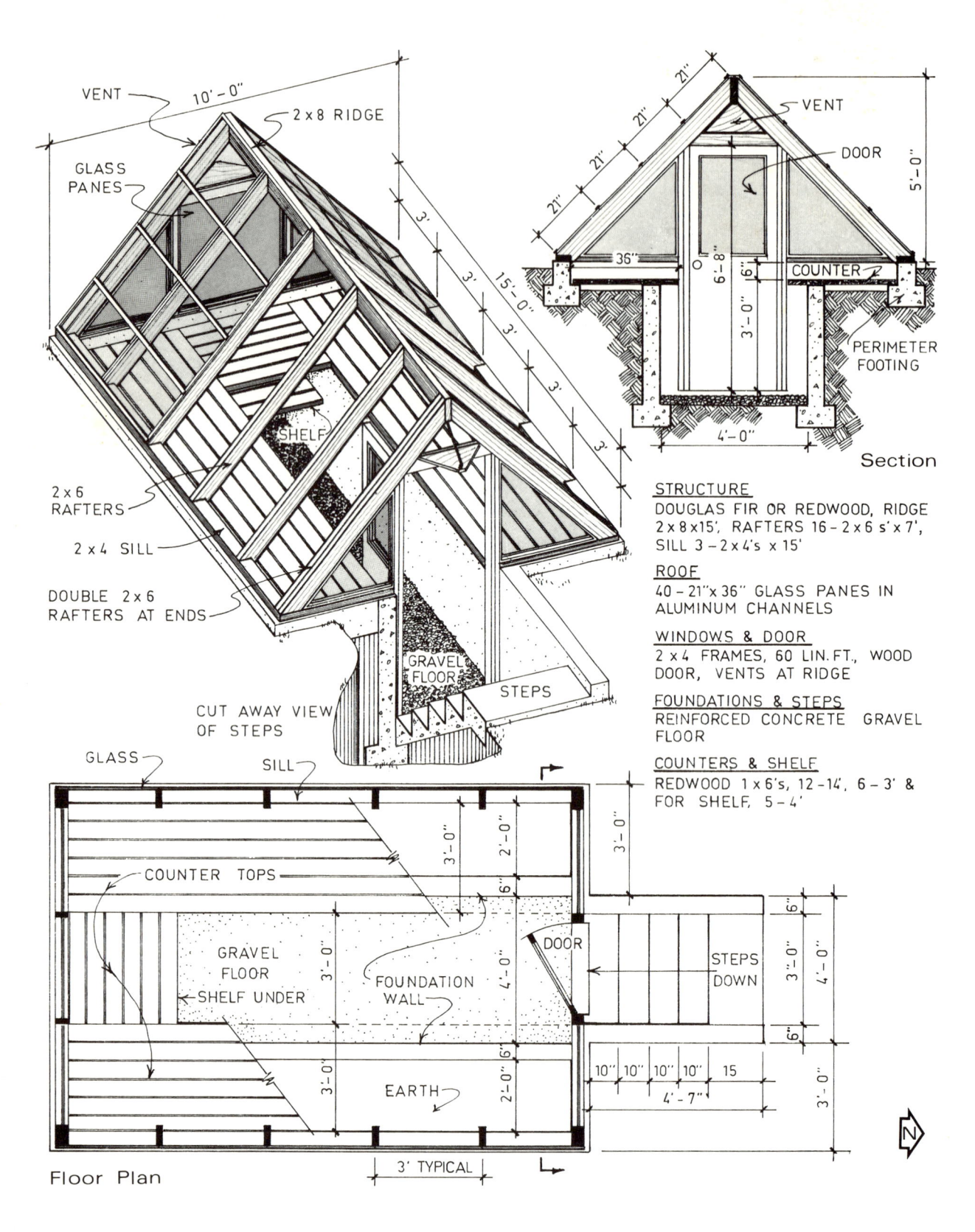

STRUCTURE
DOUGLAS FIR OR REDWOOD, RIDGE 2 x 8 x 15', RAFTERS 16 – 2 x 6's x 7', SILL 3 – 2 x 4's x 15'

ROOF
40 – 21" x 36" GLASS PANES IN ALUMINUM CHANNELS

WINDOWS & DOOR
2 x 4 FRAMES, 60 LIN. FT., WOOD DOOR, VENTS AT RIDGE

FOUNDATIONS & STEPS
REINFORCED CONCRETE GRAVEL FLOOR

COUNTERS & SHELF
REDWOOD 1 x 6's, 12 – 14', 6 – 3' & FOR SHELF, 5 – 4'

Floor Plan

Pit Greenhouse

Adrian Martinez

same things as the indoor/outdoor rooms in Part One. Namely, they use the sun's energy for heat, are geographically placed so they benefit from natural vegetation, and are built with proper materials. Whether you have professional help or are doing it yourself, there are some basic constrtuction facts you should know.

The Pit Greenhouse

Locate the greenhouse where it will benefit from the fullest sunlight. Deciduous or evergreen trees should not be near it because they will obscure the light. The site should have good drainage and not be too difficult to excavate. Digging out the pit—and this is hard work—is the first consideration. You will have to dig down at least 4 feet (the earth removed may be used for grading around the pit or disposed of).

A pit that is muddy or even slightly full of water is of no use to anyone, so drainage is a prime part of the building technique and must be considered before any building is started, not afterward. A standard method of ensuring good drainage is to install a 12-inch layer of graded stones at the bottom of the excavation. Cover this layer with some crushed stone or pea gravel to walk on. However, if the pit is on very level ground or in heavy soil, lay drainage tiles along the bottom of the excavation close to the walls and outward to lower ground to ensure absolute drainage. Run this tile outside the wall, and then cover it with loose stones (see drawings #14 and #15).

The lining of the pit or the walls may be concrete or concrete block (earth will not hold by itself). Solid concrete is poured, which means you must build or rent wood forms for the walls. Reinforce the walls with rods or wires. Set up the wood forms to shape the walls; then brace them. Pour the concrete and allow it to harden, and then remove the wood forms. Unless you are handy with tools, have this work done by a professional. (Details for concrete pouring are in Chapter 4.)

Concrete block is an easier method of wall construction. Set the blocks in place (as you would install brick row by row). To my way of thinking the concrete block wall is quite satisfactory and saves work and time: you do not have to remove the wooden forms that are necessary when using concrete. Indeed, removing concrete forms can be tricky, especially in confined areas such as a pit greenhouse.

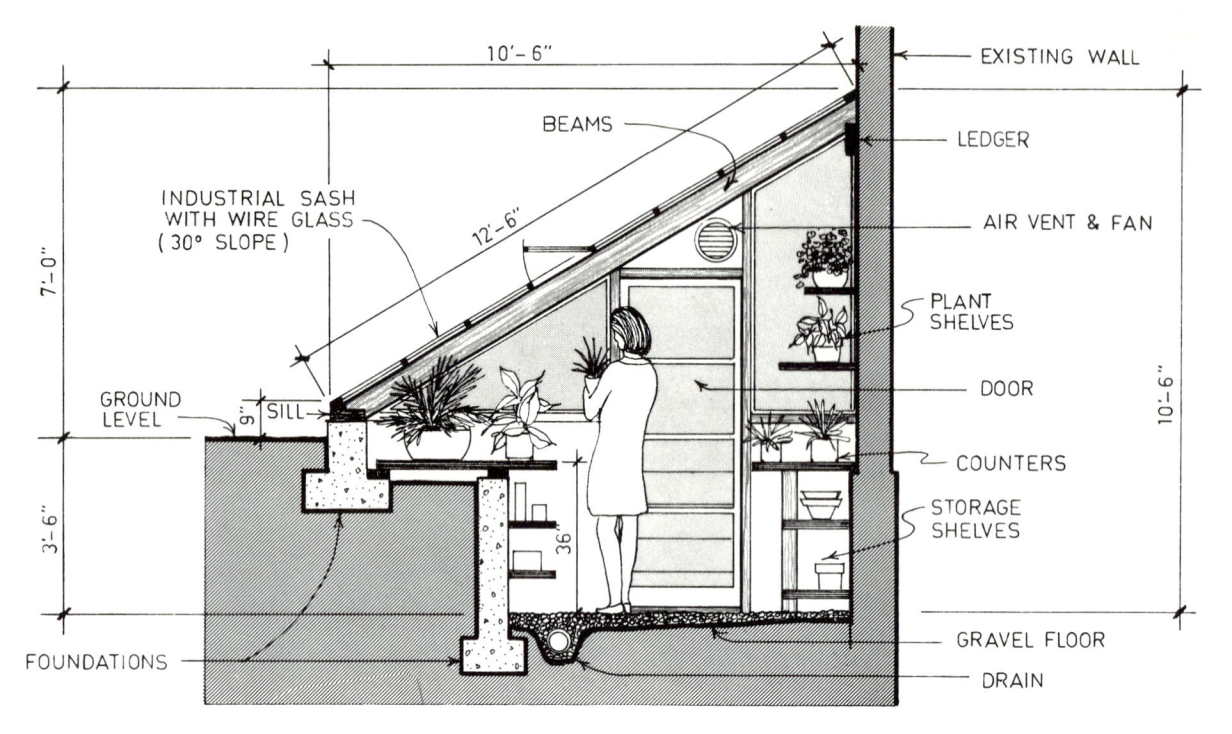

Section

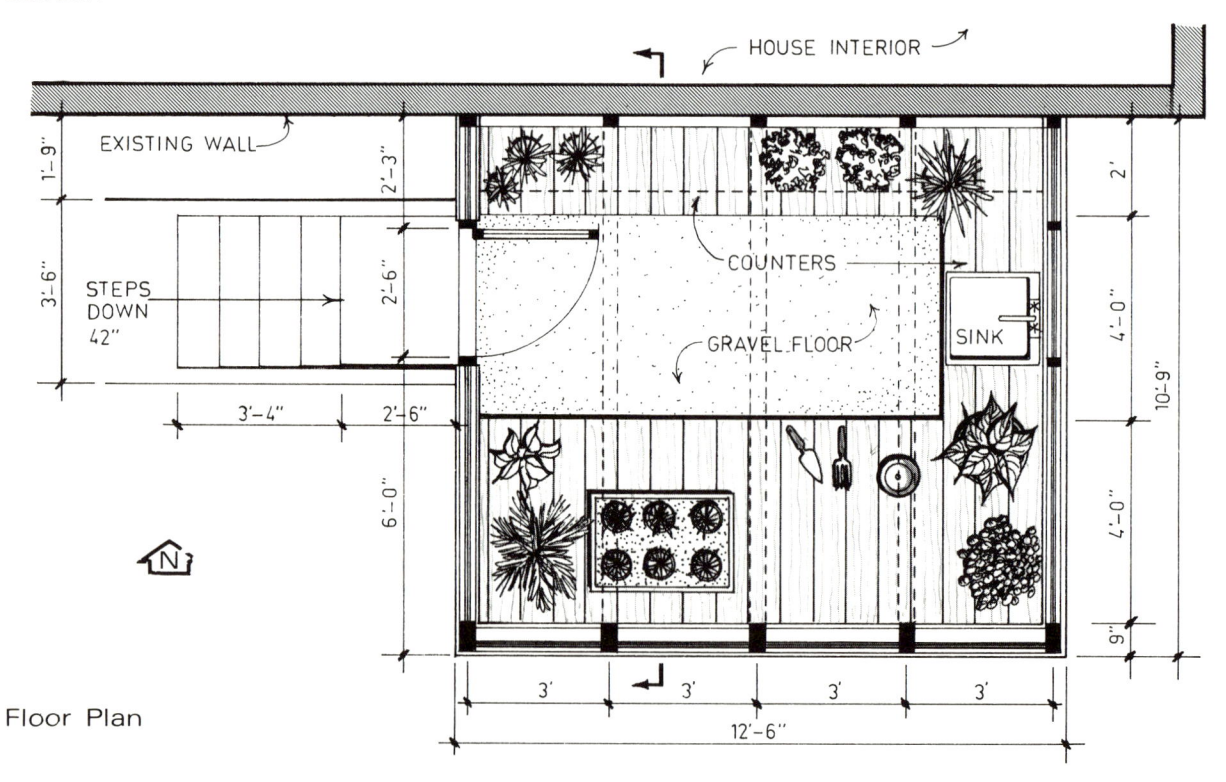

Floor Plan

Pit Greenhouse / Lean-to Adrian Martinez

STRUCTURE
4 x 6 BEAMS, 3' ON CENTRE, ON 2 x 6 LEDGER AGAINST HOUSE WALL & WOOD SILL ON CONCRETE FOUNDATION WALL, WOOD FRAMED ENDS

ROOF
STEEL INDUSTRIAL SASH, 36"x18" SUBDIVISIONS W/ WIRE GLASS, ONE ROW OPERABLE

WINDOWS
FIXED GLASS IN WOOD FRAMES

DOOR
WOOD FRAMED W/GLASS PANELS, AIR VENT & EXHAUST FAN ABOVE

FOUNDATION WALLS FOOTINGS & STEPS
REINFORCED CONCRETE, TEXTURED SURFACE ON STEPS & LANDINGS

FLOOR
SMOOTH GRAVEL OVER EARTH W/DRAINAGE TILE

COUNTERS
2 x 8 REDWOOD BOARDS ON REDWOOD STRUCTURE STAINLESS STEEL SINK, PLYWOOD CABINET BELOW SUSPENDED ABOVE FLOOR

SHELVES
1 x 6 REDWOOD BOARDS ON STEEL SUPPORTS

Pit Greenhouse / Lean-to

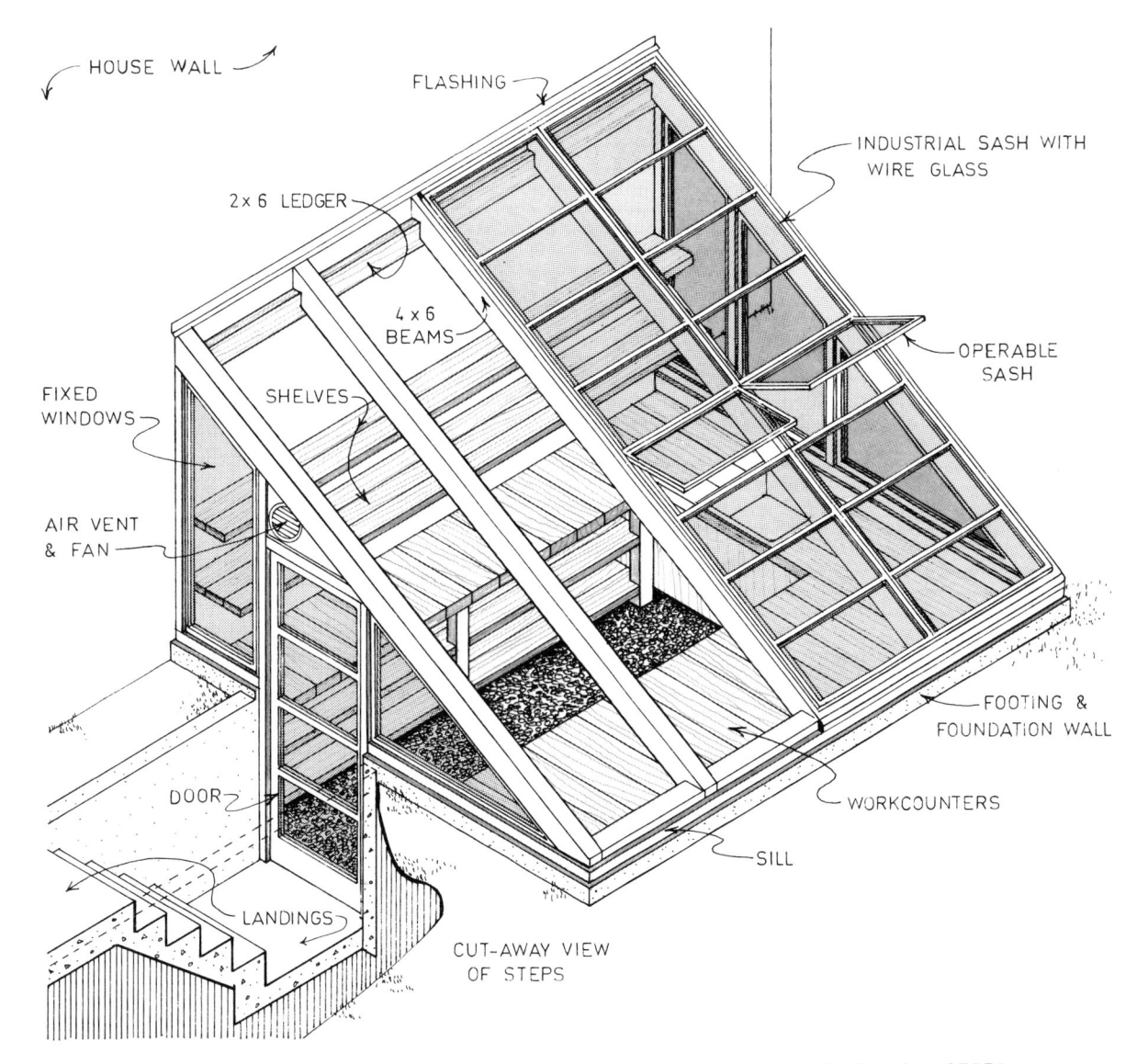

All parts of the pit above ground should be insulated with waterproof insulation: tar paper or fiberglass. The insulation is easily installed between the studs and rafters. The north side of the roof must be insulated; roofing tar paper or shingles are satisfactory. The south side will of course be glass. You can use ordinary 3- x 6-foot hotbed sash (available from suppliers), but for a tighter and more draft-free structure use a one-piece welded skylight. This is more expensive than sash but looks and performs better. In the attached pit the north roof may slope back, not to the ground, but to the building behind so the angle formed may be less than the standard 45 degrees.

Ventilation, of prime importance in the pit, can be provided by openings at both ends and movable hopper windows in the north roof. In winter these openings can be properly sealed and insulated with shutters or boarded. You will need a door at one end. The basic design of the pit allows for a small opening, perhaps a 2-foot door at the most, which could be a tight squeeze for some people, so be prepared. Use steps or some form of ladder to get down into the pit.

During the winter the south glass wall will need some protection. A sheet of canvas is suitable, or shutters can be made to fit the openings, but this means constant opening by day and closing by night unless you hinge them properly so it is only a matter of pulling them open and shutting them. Avoid lifting each one in place because it is simply too much work. A covering of hay or straw will insulate the pit glass and is the cheapest solution.

In summer the south wall will need some protection from intense heat. Use wooden roller blinds (quite inexpensive now). Inside, arrange shelf space for plants so you can easily reach pots. A 3-foot wide shelf is generally good. Use shelves and hanging devices to accommodate plants. The inside arrangement of the pit is actually dependent upon just what you are growing and how many plants you grow. By all means install a water line so there is water readily available in the pit. This can be an extension of an existing outdoor line.

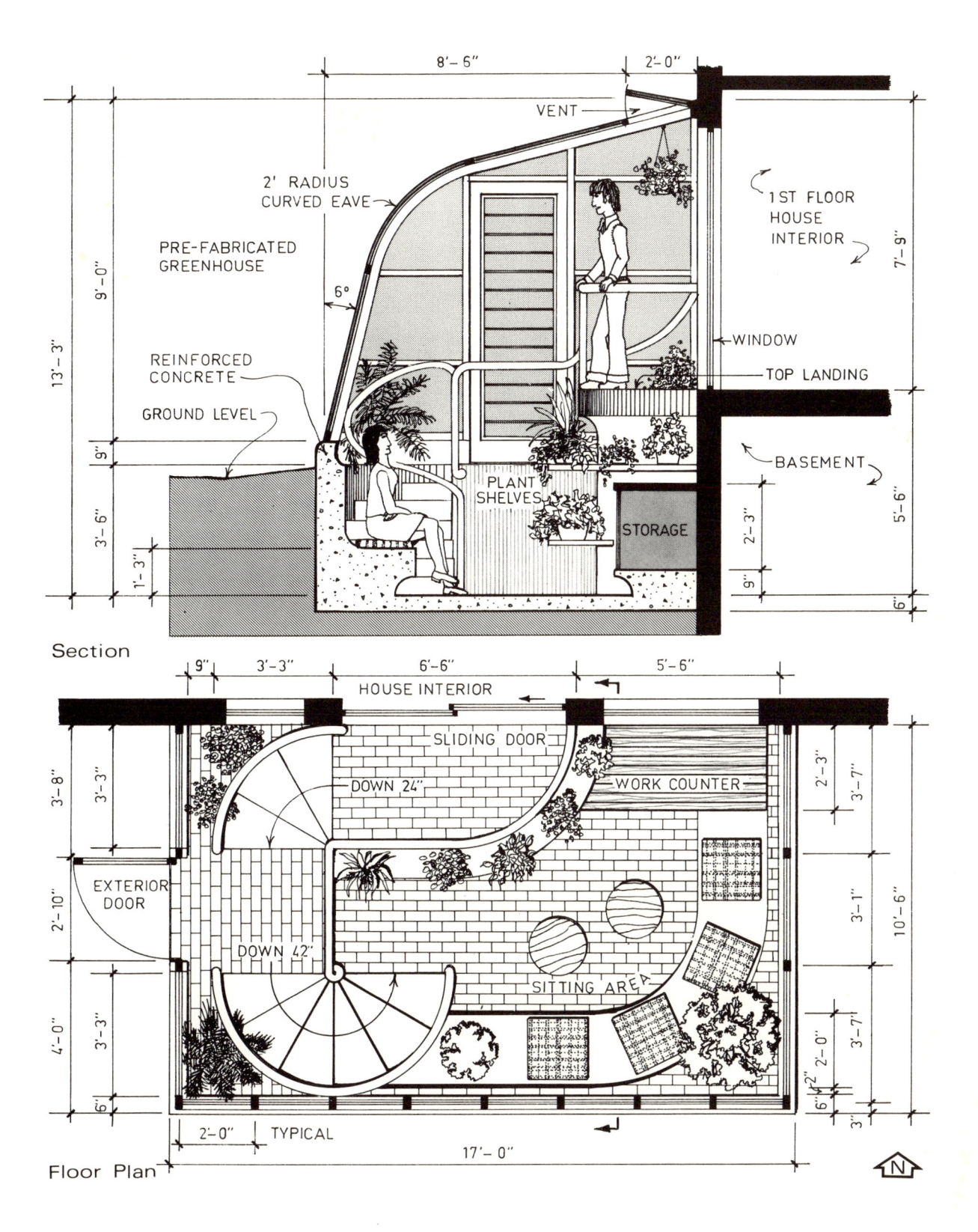

Recessed Greenhouse

Adrian Martinez

Section

Floor Plan

The Lean-to Greenhouse

The lean-to is generally above grade, with a concrete foundation. It is usually larger, and if equipped with a solar-type heating element, it can be used for plants almost all year (see Chapter 2). Auxiliary space heaters (electric) may have to be used during the coolest months. In many ways the lean-to is a modern version of the old-fashioned greenhouse, where glass sloped to the south from a wall on the north side of the home. It is economical to build but, as mentioned, will not retain as much sun heat as the pit for night heating. The actual construction, however, is more complicated. One of the major considerations, besides the foundation, is a tight joint between the wall and the roof of the home to which it is attached.

You can add a solar-type heating unit if you make your own lean-to. The simplest unit is the collector type, which is actually a hot-water system (see Chapter 2). Concrete walls to hold heat and suitable shutters to cover the sides of the greenhouse on very cold nights should be used. This greenhouse can be completely above ground as a regular structure, or part of it can be sunken, with concrete walls to absorb even more heat. Either way, a concrete foundation will be necessary.

If you make a lean-to greenhouse, use thermopane glass for the south wall even though it is more costly than regular glass because it conserves heat and saves as much as 37 percent of the heat within the structure. Thermopane comes in standard stock sizes and can be easily incorporated into the design. For the other walls, use glass and wood or concrete block or brick rather than solid glass construction. This greenhouse will function economically if built in such a manner and will be more advantageous than a prefab, all-glass house. Be sure to allow ample provision for ventilation. Houses can get very hot in summer when there is intense sun, but if ventilation is good, heat will be kept under control. Incorporate special roof vents into the design, and use windows you can open and close easily.

The Dome Greenhouse

Geodesic domes as greenhouses have become popular in the last few years because these structures are inexpensive to build and offer ample growing height for plants. They can be complicated or easy to build, depending upon the design you choose. (There are hexagons, octagons, polyhedrons, and so forth.)

No matter which dome you choose, avoid making it completely glass (or of flexible plastics). It will be too drafty and to difficult to glaze properly. Make the house small, and use thermopane or ¼-inch plexiglass in one area and insulating board or equal materials in other areas. The dome, like the lean-to, may have parts of its structure underground to conserve heat or above ground with partial concrete foundation walls. Floors may be concrete or, to save money, a cinders-on-earth can be used.

A dome made of a combination of plexiglass or thermopane and insulating board will work finely to conserve heat. This type of construction rather than a complete glass unit will be best for plants.

Some domes are built without adequate foundations or minimally on posts or piers, but the dome greenhouse should have a solid concrete foundation and preferably one solid concrete wall facing north to retain heat. For ventilation—and this is a must—have some windows that can be opened. To further ensure against heat loss in winter, partially sink the dome, following the plan of a pit greenhouse.

Structural components on the dome (skeleton) vary from aluminum to various metal channels, but for the greenhouse a redwood skeleton is the best. Although we have increased the cost factor with plexiglass and redwood, we have designed a far better, longer lasting structure than the conventional dome as it is known today.

In the dome greenhouse you have a latitude of space for plants and with some imagination can fashion a unique place for plants. Suitable benches and shelving have to be built. This is a project for ambitious people, but once done it is indeed a unique structure and well worth the time and effort, and, if done properly, will be far cheaper than any other conventional greenhouse.

Because of their design most dome greenhouses, as mentioned, will not suit the home. Placed adjacent to the home the greenhouse looks awkward and out of place. However, when connected by a glass gallery it can become a desirable addition to almost any type of house.

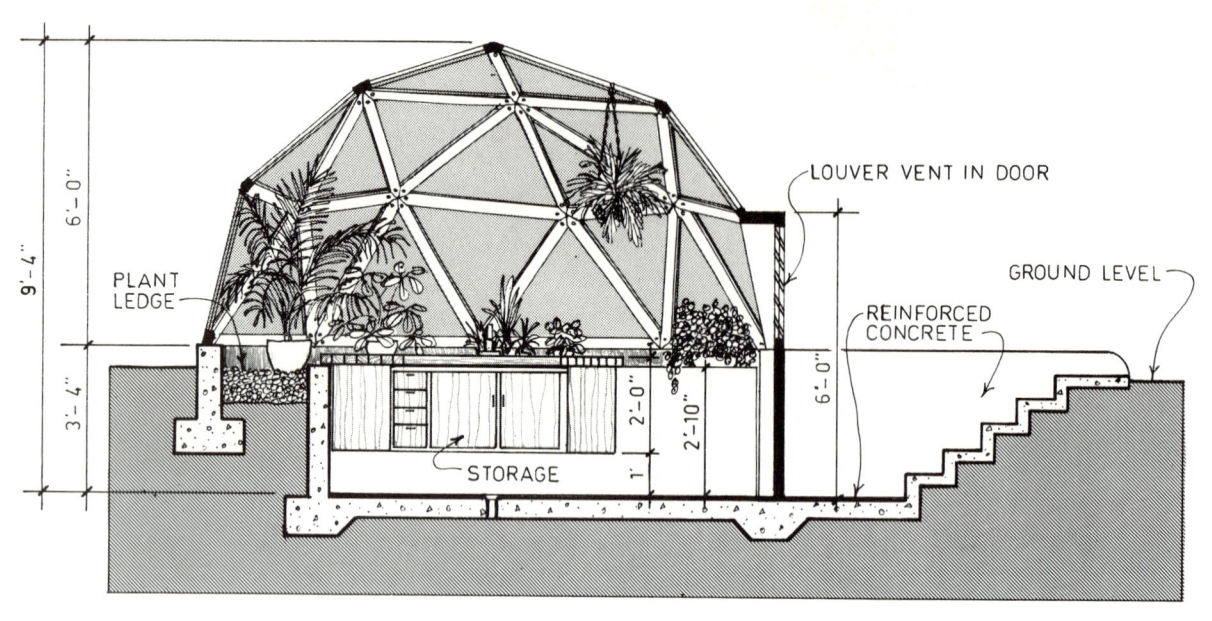

Section

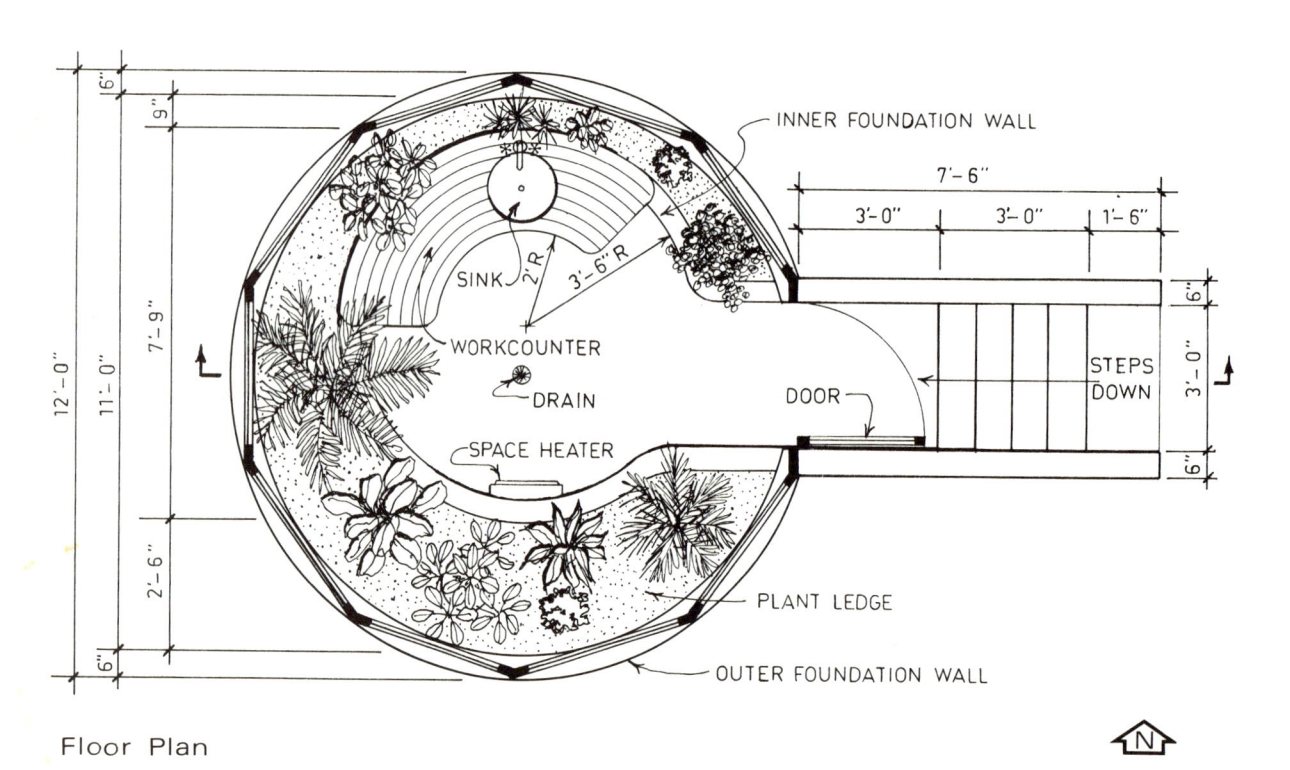

Floor Plan

Geodesic Greenhouse

Adrian Martinez

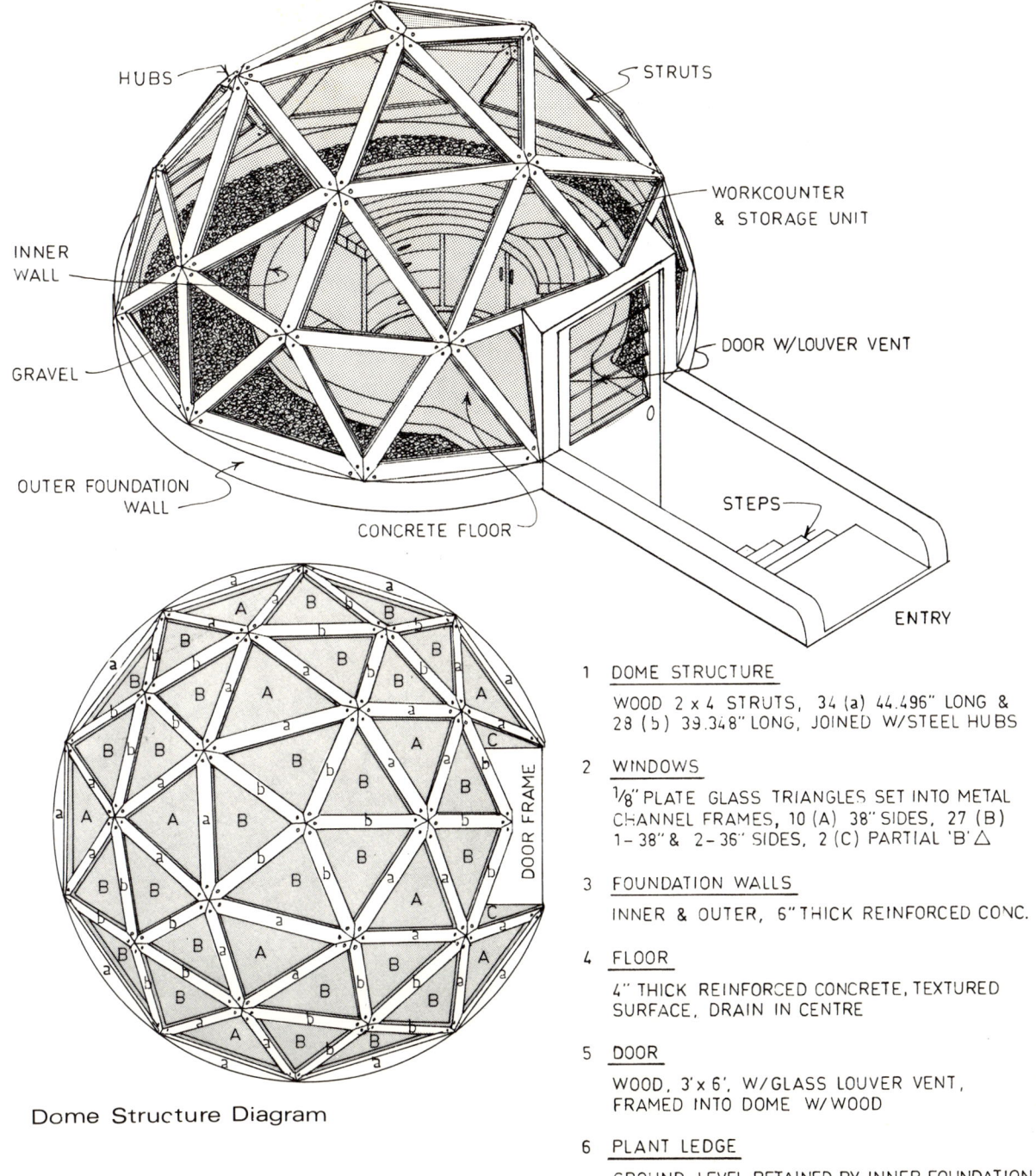

Labels on upper diagram:
- HUBS
- STRUTS
- WORKCOUNTER & STORAGE UNIT
- INNER WALL
- DOOR W/LOUVER VENT
- GRAVEL
- STEPS
- OUTER FOUNDATION WALL
- CONCRETE FLOOR
- ENTRY

Dome Structure Diagram

NOTE:

THIS IS A GEODESIC HEMISPHERE, 2-FREQUENCY ICOSAHEDRON ALTERNATE, 12' DIAMETER

HEATED BY A WALL MOUNTED SPACE HEATER WITH A THERMOSTAT

1. **DOME STRUCTURE**

   WOOD 2 x 4 STRUTS, 34 (a) 44.496" LONG & 28 (b) 39.348" LONG, JOINED W/STEEL HUBS

2. **WINDOWS**

   1/8" PLATE GLASS TRIANGLES SET INTO METAL CHANNEL FRAMES, 10 (A) 38" SIDES, 27 (B) 1-38" & 2-36" SIDES, 2 (C) PARTIAL 'B' △

3. **FOUNDATION WALLS**

   INNER & OUTER, 6" THICK REINFORCED CONC.

4. **FLOOR**

   4" THICK REINFORCED CONCRETE, TEXTURED SURFACE, DRAIN IN CENTRE

5. **DOOR**

   WOOD, 3' x 6', W/GLASS LOUVER VENT, FRAMED INTO DOME W/WOOD

6. **PLANT LEDGE**

   GROUND LEVEL RETAINED BY INNER FOUNDATION WALL, SURFACED W/GRAVEL FOR DRAINAGE

7. **WORK/STORAGE UNIT**

   LAMINATED WOOD COUNTER W/SINK, WALL HUNG PLYWOOD CABINET

Geodesic Greenhouse

Foundations

Whatever greenhouse you select, attached or free-standing, it should rest on a foundation. (An exception may be the geodesic structure in all year temperate climates.) This can be poured concrete, steel, or redwood. The footings extend from the ground level to a few inches above the grade, down to frost level or a few inches below the frost line. Frost lines must be observed for footings because alternating freezing and thawing of the ground will cause walls to crack, and building codes require that frost lines be observed. Call your local building department to find out the frost line in your area. As mentioned previously, greenhouse prices do not include foundation or footings.

If you do not want to hire a professional to do all the greenhouse construction, have him dig out and pour the footings; you can install the foundation wall. This is not difficult. Use premixed mortar and prepare a little at a time. Add water to the mix, and with a hoe work the moisture throughout the mortar to get a heavy pastelike consistency. If you are having a brick foundation wall, lay the bricks moist, never dry. Put mortar on the footings for the first course or row of bricks. Using the trowel upside down, press ripples into the mortar to provide a good gripping surface for the bricks. Remove excess mortar with a trowel. Lay brick number one for the second row, and use the trimmed mortar to smear the edge of the next brick. Be sure to keep the rows of brick even and level. Remove all loose bits of mortar from the wall before they have a chance to set. Cover the wall with wet burlap for a few days to permit a slow cure and to prevent cracks. A few weeks later clean the walls with muriatic acid, then hose them down thoroughly. (Do not use mortar when temperature is below 40°F; it will cause cracking in the wall.) In mild climates greenhouses can be installed on poured concrete slabs.

Although a dirt floor covered with gravel or crushed stones is adequate, a brick or tile floor is more attractive. Or use stepping stones (patio blocks) embedded in gravel. These materials retain moisture, and evaporation of water on the floor creates humidity in the greenhouse. Construction details including glass and glazing are the same as those discussed in Chapter 4, the construction of the indoor/outdoor room.

8 The Greenhouse For You

JUST WHAT KIND OF GREENHOUSE YOU CHOOSE DEPENDS UPON your individual climate, the site, and the kind of soil. In all-year temperate climate the problem is not how much artificial heat for evenings (since they are mild), but how to keep the growing area cool in summer. In severe winter areas, the problem is reversed. In regions where there are extremes in heat and cold, still other methods must be employed to arrive at the most functional greenhouse.

The site, as mentioned earlier in this book, is also important. Placing the greenhouse where it looks well and has natural protection yet free access to south sun must be considered.

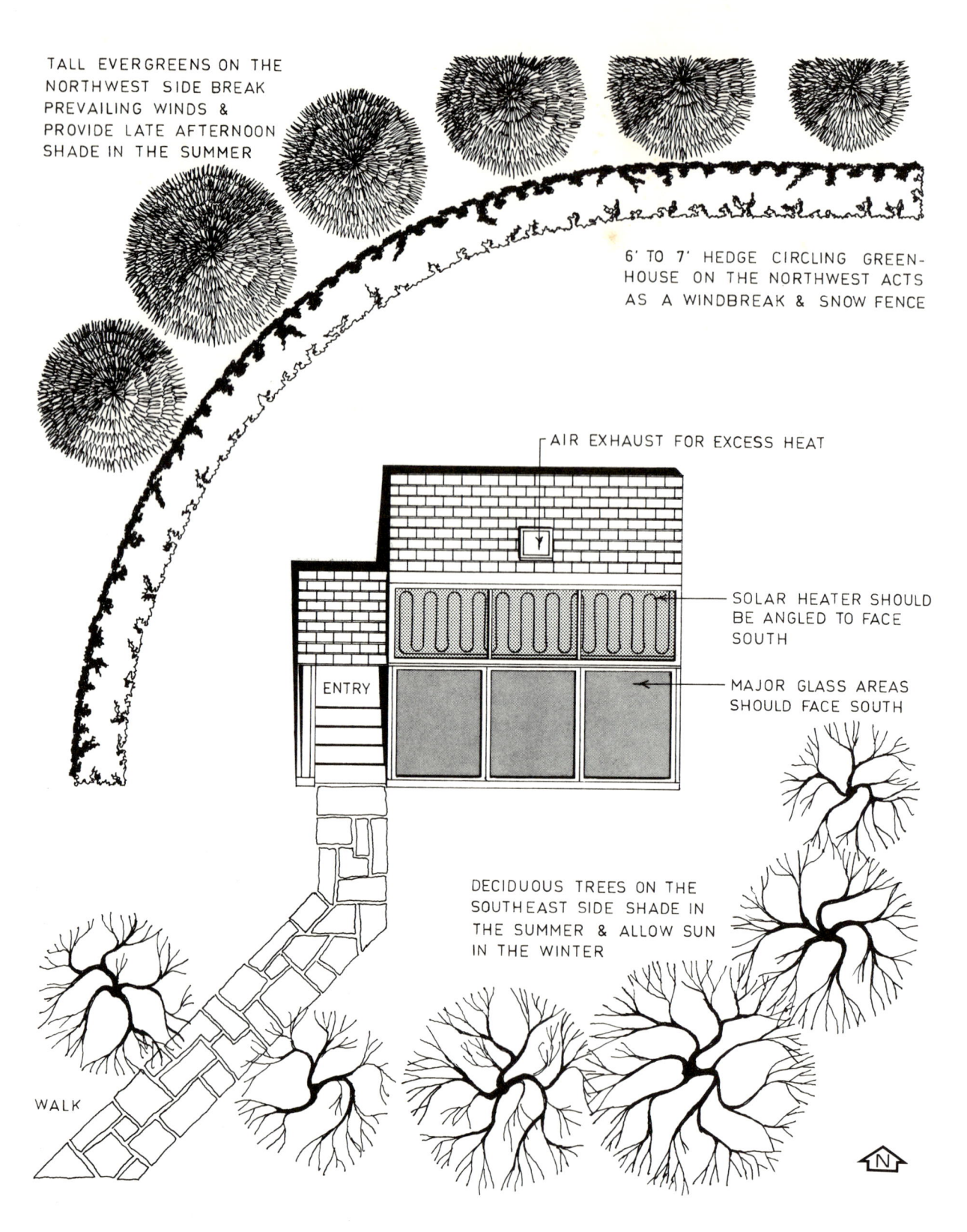

TALL EVERGREENS ON THE NORTHWEST SIDE BREAK PREVAILING WINDS & PROVIDE LATE AFTERNOON SHADE IN THE SUMMER

6' TO 7' HEDGE CIRCLING GREENHOUSE ON THE NORTHWEST ACTS AS A WINDBREAK & SNOW FENCE

AIR EXHAUST FOR EXCESS HEAT

SOLAR HEATER SHOULD BE ANGLED TO FACE SOUTH

MAJOR GLASS AREAS SHOULD FACE SOUTH

ENTRY

DECIDUOUS TREES ON THE SOUTHEAST SIDE SHADE IN THE SUMMER & ALLOW SUN IN THE WINTER

WALK

Cold Climate Site Plan

Adrian Martinez

For Temperate Climates

Air conditioning is the usual answer to how to have cool space in hot climate areas, but this involves equipment and expensive operating costs. With air conditioning it costs three to five times as much to remove a BTU of heat from a space as it does to add one in winter. There are other ways to cool the greenhouse, natural ways. Roofs can become unbearably hot in direct sun, reaching temperatures of 150°F. You think the obvious answer is insulation, but this is not so because a light frame construction heats rapidly during the day and thus cools better at night. Avoid heavily constructed buildings in hot areas.

Roof pitch in the greenhouse greatly influences the heat factor. A perfectly flat roof adds 50 percent more heat gain than a pitched roof. It is difficult to get a natural hot air flow out of a flat space but easy with a pitched roof. Heat flows through building materials by convection, radiation, and conduction, and it is radiation that is most significant in summer heating. Metallic surfaces diminish heat radiation by simply bouncing them off, whereas materials like stone or brick gather heat. Reflective metal surfaces on the roof (shiny side out) can contribute greatly to diminishing heat. (There should be a ventilated air space inside or below of ¾ inch between the reflective surface and the roof.) A white or light-colored roof will bounce off about 70 percent of the sun's heat rays. In very, very hot climates a greenhouse roof with white gravel over tar is most effective. The interior will remain cool for a long time into the night. Also to be considered, and a good idea, is to use an elevated structure—Douglas fir poles, for example—so air can circulate under the greenhouse. The elevated building has many advantages in warm climates, so do consider it.

An exhaust or ventilating fan (easy to install and operate) combined with a tar and gravel roof can do as much to cool a greenhouse as air conditioning. And this is exactly what should be used in all-year temperate climates. This coupled with proper venting systems solves excessive heat in the greenhouse of hot regions. And with a sufficient overhang on the south side to protect the room from intense summer sun, yet allow it to enter in winter, you have built the best possible greenhouse for your area. See Drawing #21.

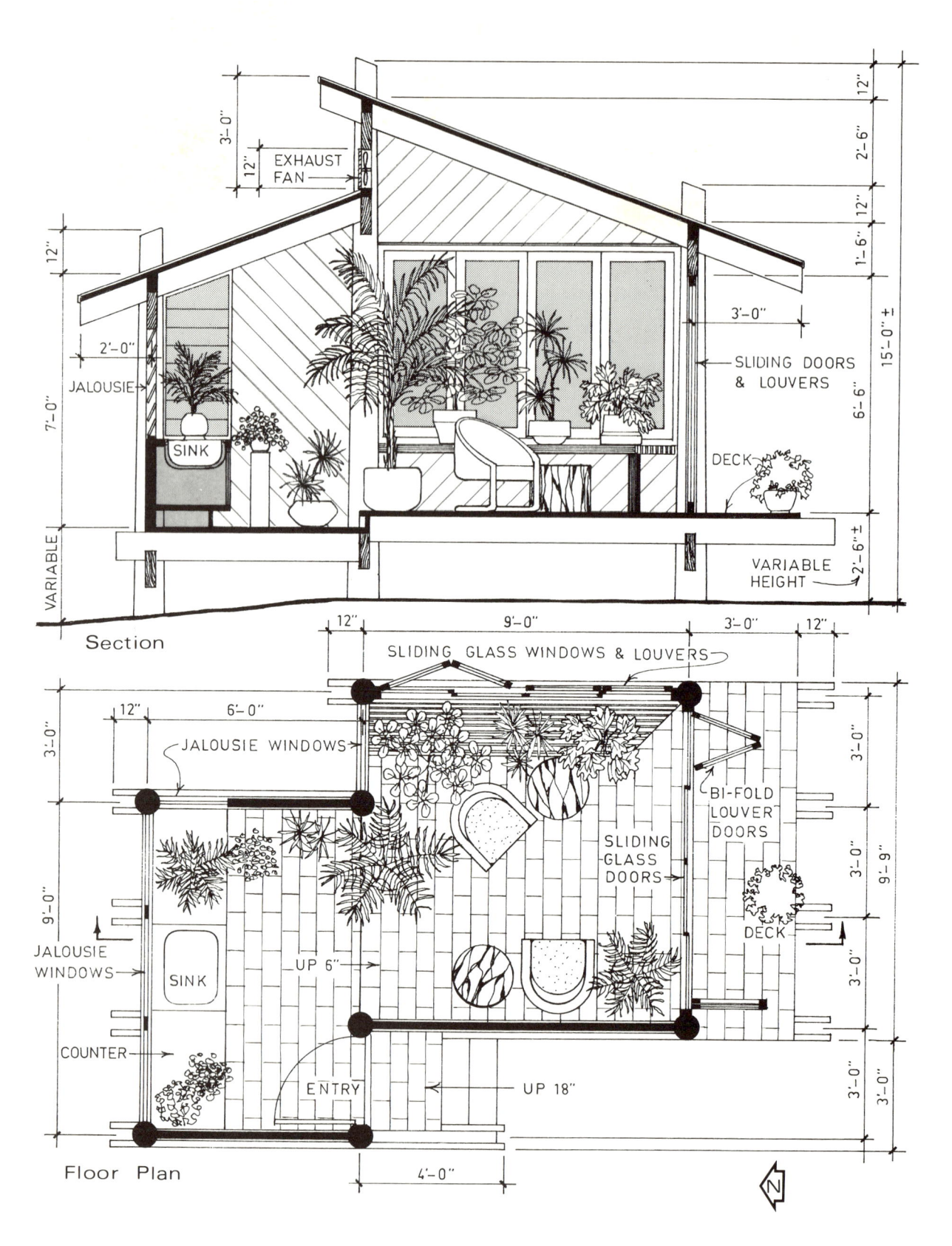

Hot Climate Greenhouse

Adrian Martinez

Section

Floor Plan

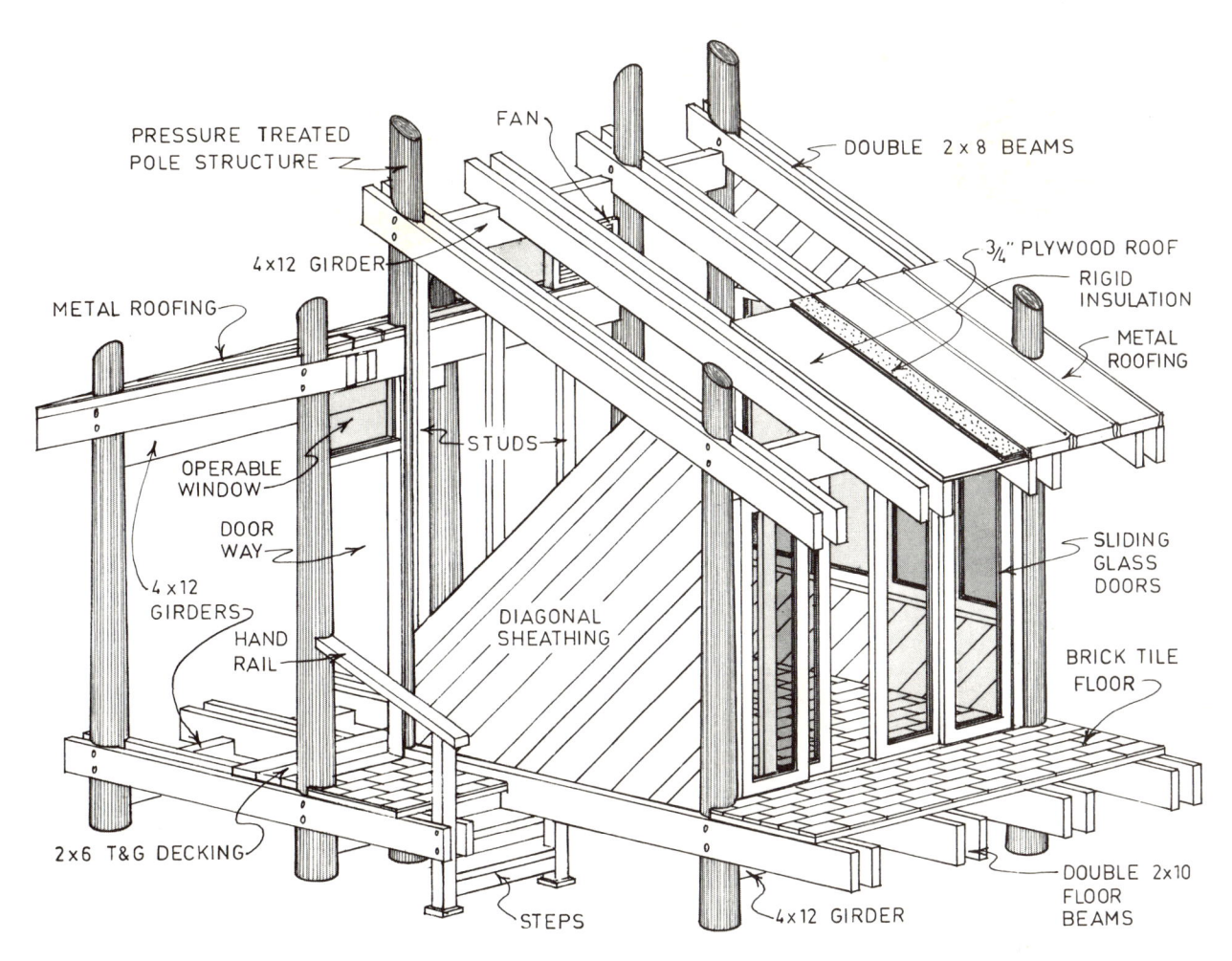

STRUCTURE
9" DIA. DOUGLAS FIR POLES, PRESSURE TREATED WITH A PRESERVATIVE. FOUNDATIONS, IF REQUIRED, DEPEND UPON SOIL CONDITIONS

ROOF
DOUBLE 2 x 8 BEAMS BOLTED TO POLES OR SUPPORTED BY 4 x 12 GIRDERS, 3/4" PLYWOOD ROOF DECKING, RIGID INSULATION & LIGHT COLORED METAL SHEATHING

FLOOR
DOUBLE 2 x 10 BEAMS BOLTED TO POLES OR SUPPORTED BY 4 x 12 GIRDERS, 2 x 6 TONGUE & GROOVE FLOOR DECKING, SEALED & SURFACED WITH BRICK TILE

WALLS
WOOD FRAMED WITH BLANKET INSULATION BETWEEN STUDS & 1" RANDOM WIDTH DIAGONAL SHEATHING

WINDOWS
GLASS JALOUSIE; WOOD FRAMED SLIDING WITH WOOD LOUVERED SHUTTERS; & OPERABLE METAL FRAMED CLERESTORIES

WORK COUNTER
WOOD CABINETS WITH LAMINATED WOOD COUNTER TOP & STAINLESS STEEL SINK

DOORS
28" x 6'-8" WOOD AT ENTRY & SLIDING WOOD FRAMED GLASS WITH BI-FOLD LOUVERED WOOD SHUTTERS

NOTE: STRUCTURE IS ELEVATED TO ALLOW FOR AIR CIRCULATION, SOUTH SIDE IS SHADED BY ROOF OVERHANG, GLASS ON EAST & SOUTH SIDES IS PROTECTED BY SHUTTERS, WEST WALLS ARE WINDOWLESS & INSULATED, ROOF LIGHT COLORED TO REFLECT HEAT

Hot Climate Greenhouse

In a temperate all year climate, plastic sheet (rigid) can be used for a greenhouse. The floor is earth and the room is protected on two sides by natural vegetation. (Photo by Matthew Barr).

This green house in a temperate-all-year climate also uses plastic sheet with a concrete block foundation wall. (Photo by Matthew Barr).

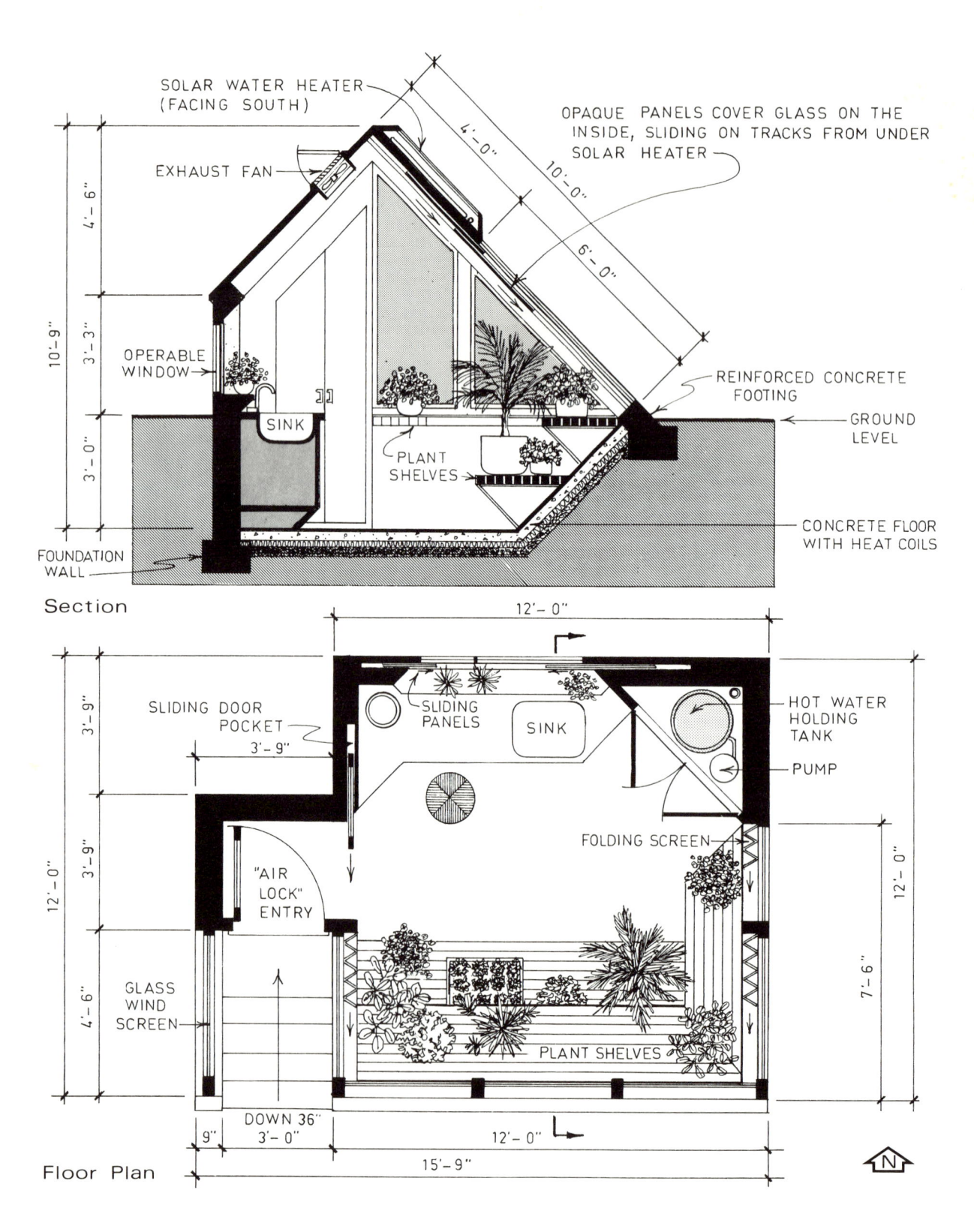

Cold Climate Greenhouse

Adrian Martinez

Section

- SOLAR WATER HEATER (FACING SOUTH)
- EXHAUST FAN
- OPAQUE PANELS COVER GLASS ON THE INSIDE, SLIDING ON TRACKS FROM UNDER SOLAR HEATER
- OPERABLE WINDOW
- SINK
- PLANT SHELVES
- REINFORCED CONCRETE FOOTING
- GROUND LEVEL
- CONCRETE FLOOR WITH HEAT COILS
- FOUNDATION WALL

Dimensions: 4'-0", 10'-0", 6'-0", 4'-6", 3'-3", 10'-9", 3'-0"

Floor Plan

- SLIDING DOOR POCKET
- SLIDING PANELS
- SINK
- HOT WATER HOLDING TANK
- PUMP
- FOLDING SCREEN
- "AIR LOCK" ENTRY
- GLASS WIND SCREEN
- PLANT SHELVES
- DOWN 36"

Dimensions: 12'-0", 3'-9", 3'-9", 4'-6", 12'-0", 7'-6", 9", 3'-0", 15'-9"

N

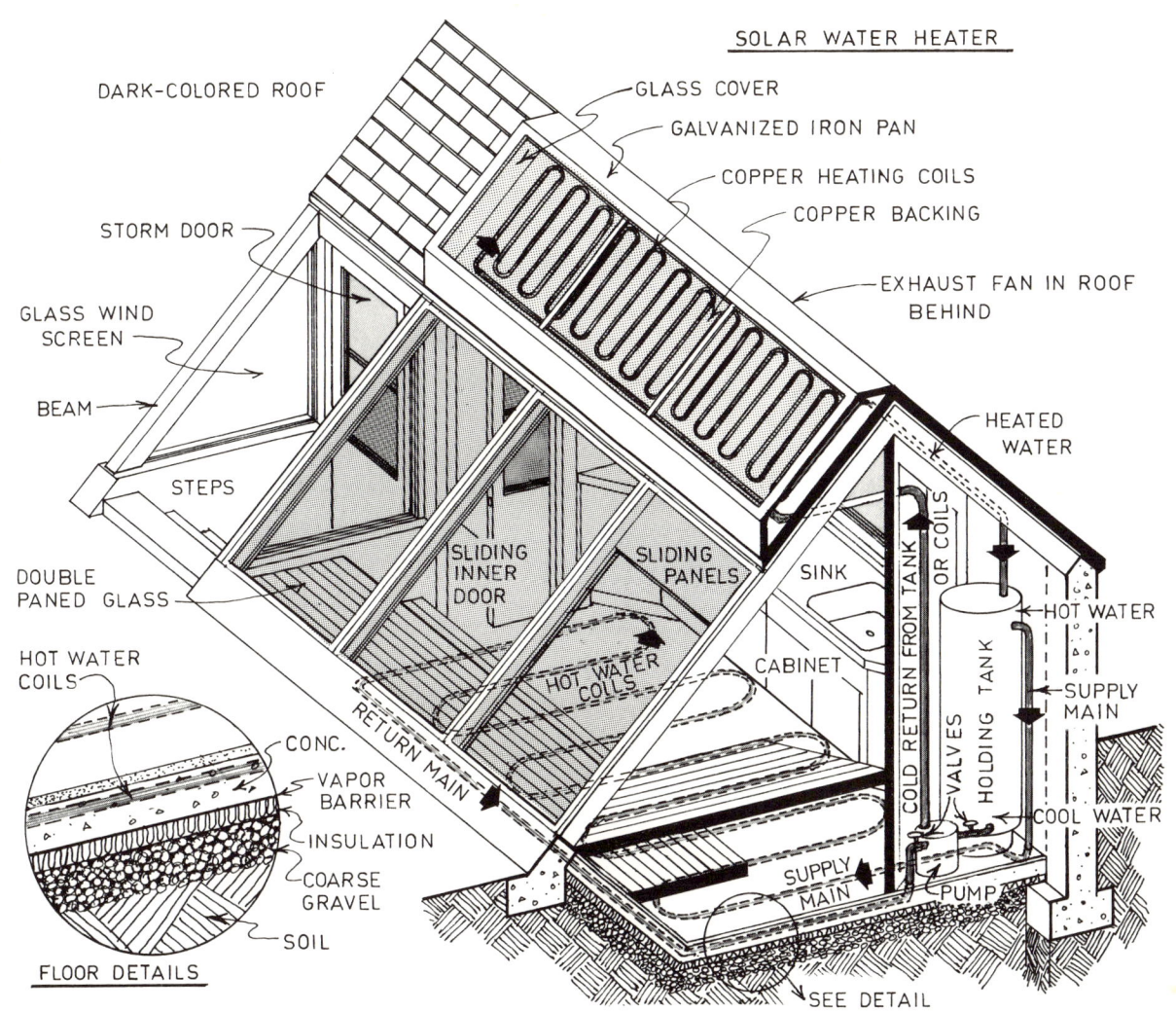

ROOF
4 x 8 DOUGLAS FIR BEAMS, ¾" PLYWOOD DECKING, RIGID INSULATION, BUILT-UP ROOFING, DARK SHINGLES

WALLS
9" THICK, SEALED, REINFORCED CONCRETE

FLOOR
4" REINFORCED CONCRETE, VAPOR BARRIER, INSULATION & GRAVEL UNDERNEATH. DARK-COLORED CONCRETE SURFACE

WINDOWS
METAL FRAMED, DOUBLE-PANED GLASS, WINDOWS OVER COUNTER & IN DOORS OPERABLE

DOORS
INSULATED STORM DOOR OUTSIDE & SLIDING DOOR INSIDE, FORMING AN AIR LOCK

WORK COUNTER & SHELVES
WOOD CABINET & SHELVES TILED COUNTER WITH STAINLESS STEEL SINK

SOLAR HEATER
AS ILLUSTRATED ABOVE. WATER IS HEATED BY THE SUN, STORED IN THE HOLDING TANK, THEN PUMPED & CIRCULATED THROUGH FLOOR COILS TO HEAT THE FLOOR AND RADIATE INTO THE GREENHOUSE. COOL WATER RETURNING FROM FLOOR COILS CAN BE PUMPED DIRECTLY TO HEATER OR WHEN SYSTEM IS OFF, COOL WATER IN TANK CAN BE PUMPED TO HEATER KEEPING WATER IN TANK HOT.

NOTE: STRUCTURE IS SUNKEN TO RETAIN HEAT, WALLS ARE THICK, FLOOR & ROOF INSULATED. DOUBLE PANED GLASS WINDOWS HAVE SLIDING PANELS OR SHUTTERS TO RETAIN OR REPEL HEAT, AS REQUIRED. SPACE IS VENTED.

Cold Climate Greenhouse

For Cold Climates

In winter climates, building materials that retain heat—stone, brick—should be used. (This is why prefabricated greenhouses of all glass and metal are so ineffective in these areas.) The greenhouse that faces south is a must. In winter the sun's rays warm the space, but in summer the path of the sun is overhead so there is needed shade. Thus the south glass wall with an overhead for the greenhouse and a lower wall on the north side to keep out cold winds is basic good building. Storing the surplus daytime heat should also be considered so it can be released at night or on no-sun days. The new type of solar collector now being manufactured can be used to get the heated water stored in a tank to circulate on cloudy days through pipes in the greenhouse walls or floors.

Walls should be concrete and a dark colored reinforced concrete floor is an excellent idea. Use double glazed windows where possible and insulated doors. If it is feasible sink the structure partially in the ground (as with the pit) to retain heat. Be sure floors and roofs are well insulated. Use wind screens where necessary. A greenhouse such as the one in Drawings #21 and #22 will require no artificial heat since the solar collector is used.

Site

When placing your greenhouse try to select a level area. This is the easiest building space. Drainage problems will be very difficult to remedy if the building is in a gully or valley where rain water accumulates. On the other hand, when placed on hills the building is too exposed to weather on one side. The trick is to use the hill to your advantage if it is a south sloping hill; that is, the shelf for the site would be on the south side of the hill, thus affording a natural barrier to weather from the north. When working on level ground you can use earth as a natural insulation. On level land check to determine water table averages in winter (call the state soil department). Never go below the mean high. In very hilly areas consider building on piers with well-insulated flooring, such as concrete. If soil is gravely or sandy, a more extensive foundation is required because in heavy clay more circulating drainage facilities are necessary.

In severe winter climates foundations must go to frost line and this greenhouse uses heavy masonry walls. It is also partially sunken to conserve on heat loss. Note the lovely summerhouse-type room next to it. (Photo courtesy Aluminum Greenhouses).

Planting

In all climates some provision should be made to use natural vegetation. In severe winters, wind and snow can be hazards, so some modification of the weather has to be considered, in addition to proper building materials, construction, and placement. This is where planting design comes in. In a forest the sun's heat is greatly reduced by trees. (If you have ever walked into a forest in summer you noticed this difference immediately.) They are cooler than cleared land in summer. Planting provides definite summer cooling. Also, windbreaks of hedges reduce heat loss from buildings by keeping out cold winds. Snow is also thwarted by a well-planted area around the greenhouse.

To demonstrate how important planting is, consider shrubs placed close to a house. In this situation the shrubs prevent breezes from penetrating the greenhouse. On the other hand, trees and grass allow heavier cool air to flow inside. Leaves and grass absorb solar radiation, and the evaporation cools the surrounding air, giving a cool flow whenever there are breezes. Also remember that of the two kinds of trees, deciduous and evergreen, the deciduous lose their leaves, so in winter the sun can penetrate through leafless branches to warm the structure. In summer they have their leaves to protect against the sun's heat.

Trees to the west of the greenhouse will give complete shade coverage; they should be staggered so the setting sun lengthens the shadows. In early afternoon the trees will protect the greenhouse from sun.

Here is a list of trees and shrubs that can be used as natural hedges against winds and storms:

List of Trees

BOTANICAL AND COMMON NAME	APPROX. HEIGHT IN FT	MINIMUM NIGHT TEMP.	REMARKS
Acer ginnala (amur maple)	20	−50 to −35F	Red fall color
Crataegus mollis (downy hawthorn)	30	−20 to −10F	Pear-shaped red fruit
Eucalyptus camaldulensis (red gum)	80–100	20 to 30F	Fine landscape tree
E. globulus (blue gum)	200	20 to 30F	Good windbreak
E. polyanthemos (silver dollar gum)	20–60	10 to 20F	Fine landscape tree
Fagus grandifolia (American beech)	120	−35 to −20F	Stellar tree
F. sylvatica (European beech)	100	−20 to −10F	Several varieties
Fraxinus americana (white ash)	120	−35 to −20F	Grows in almost any soil
Pinus nigra (Austrian pine)	90	−20 to −10F	Fast-growing tree
P. thunbergii (Japanese black pine)	90	−20 to −10F	Dense-spreading tree
Populus alba (white popular)	90	−35 to −20F	Wide-spreading tree
Quercus palustris (pin oak)	120	−20 to −10F	Beautiful pyramid
Thuja occidentalis	65	−50 to −35F	Sometimes needles turn brown in winter
Tilia americana (American linden)	90	−50 to −35F	Fragrant white flowers in July
T. cordata (small-leaved linden)	60	−35 to −20F	Dense habit
T. tomentosa (silver linden)	80	−20 to −10F	Beautiful specimen tree
Tsuga canadensis (hemlock)	75	−35 to −20F	Many uses: hedges, screens, landscape

List of Shrubs

SE = semievergreen; D = deciduous; E = evergreen

BOTANICAL AND COMMON NAME	SE D E	APPROX. HEIGHT IN FT	AVERAGE TEMP.	REMARKS
Elaeagnus angustifolia (Russian olive)	D	20	−50 to −35F	Fragrant flowers
Hamamelis vernalis (spring witch hazel)	D	10	−10 to −5F	Early spring blooms
Lagerstroemia indica (crape myrtle)	D	20	5 to 10F	Popular summer bloom
Laurus nobilis (sweet bay)	E	30	−5 to 5F	Tough plant
Lonicera tatarica (Tatarian honeysuckle)	D	10	−35 to −20F	Small pink flowers in late spring
Spiraea veitchii	D	12	−10 to −5F	Good background; graceful one
Syringa villosa (late lilac)	D	9	−50 to −35F	Dense, upright habit
S. vulgaris (common lilac)	D	20	−35 to −20F	Many varieties
Viburnum dentatum (arrowwood)	D	15	−50 to −35F	Red fall color
V. dilatatum (linden viburnum)	D	9	−10 to −5F	Colorful red fruit
V. opulus (European cranberry bush)	D	12	−35 to −20F	Good many varieties
V. prunifolium (black haw)	D	15	−35 to −20F	Good specimen plant

9 Planning and Planting

NO MATTER WHAT KIND OF GREENHOUSE YOU HAVE, YOU WILL need places for plants. In the pit greenhouse, placement of shelves and tables is vital because space is limited. The geodesic dome requires special hardware too to accommodate plants, and the lean-to greenhouse requires benches and accessories to make it function properly.

In addition to places for plants you will need adequate ventilation in the growing areas as well as humidity and auxiliary space heating when the sun does not shine. You will also need shading when the sun shines too much in summer and shutters and panels for winter protection when it gets severely cold.

Benches, Shelves, Hangers

In your greenhouse you will need places for plants. It is a good idea to build them yourself so you can use all available under-glass space. (Commercial benches are available from prefab greenhouse manufacturers.) Each structure—pit greenhouse, lean-to, or dome—will dictate just what kind of staging or benches to have. The main thing is that they are adequately supported and at proper height so you do not have to bend or squat to get to the plants. Most plant benches are made of redwood or cypress because such lumber best resists decay and moisture. Concrete benches are sometimes used, but they are heavy and cumbersome and need special forms to be cast. Never make benches more than 30 inches deep or they will be difficult to manage because plants in the rear will be in awkward positions, more than an arm's length away.

Supports for benches can be 4 x 4 redwood posts or iron pipe and should be securely anchored to the ground. A bench 24 inches wide by 60 inches long filled with soil can be very heavy. Support posts with concrete footings for best results.

Some people plant directly into the bench, that is, they fill it with soil and put plants in place, but I have found it more esthetically pleasing and much more handy to simply put plants on the bench. With this method use slatted benches: boards with ½-inch space between them so air can circulate at the bottom of pots. This also discourages insects like snails and slugs from hiding under pots. The slatted bench also provides natural drainage so excess water can run through the boards, thus preventing harmful stagnant soil.

In the pit greenhouse all you will need are some slatted shelves or benches on raised piers (4 inches on each end) on each wall. The benches are actually already there because you have dug down 4 or 5 feet to waist level. Lean-to greenhouses need full benches. Dome structures can accommodate many short benches in relation to the sides of the greenhouse. If it is a hexagonal design, six benches would be appropriate, or you might try a circular-designed bench made from exterior plywood cut and shaped to correspond to the design of the dome.

Shelves at windows are still another way to have room for plants. These may be of glass, with an attaching device, or redwood slats set ½ inch apart on brackets. I used these exclusively in one greenhouse and they were quite satisfactory. Be sure to make the shelves at least 8 inches wide so they can accommodate large as well as small pots.

Good bench space for plants is an integral part of a successful greenhouse. Here the shelves are arranged stair-step for easy access to plants. (Photo by Matthew Barr).

Hanging hardware to support pots and tubs of plants is still another way to gain space for plants and at the same time make the greenhouse handsome. A simple installation is to put eye-hooks in the wood beams of the greenhouse and then use chain from the hooks to hold plants. Or use the popular plant hangers of metal, macramé, or rope. Always remember to hang plants so they are out of the way of traffic and will not hit someone in the head.

Again what structures you use for plants depends upon the type of greenhouse you have. You may want to forego the benches because they do not, most of the time, look attractive, and just have some shelves at the windows and some hanging plants. We show some bench and shelf designs in drawings so you can have some idea of how to proceed to have places for plants.

A work and storage area such as this is a good idea in a greenhouse; it keeps everything in its place. (Photo by Matthew Barr).

Hardware Accessories

You can get very fancy and have very elaborate controls for opening and shutting windows, humidifiers, and so forth. Actually you do not need any of this. Just be sure that no matter what kind of greenhouse you have, there are windows at each end which can be opened to admit a good circulation of air. It is a simple matter to open the windows on hot stuffy days and close them when it is too cold. Two windows will be adequate for the average greenhouse.

Many plants growing together create humidity (they give off moisture through the leaves), so do not invest in fancy humidifiers because you will not need them. The humidity in your greenhouse will average about 50 to 60 percent if you have a few dozen plants, which is fine for most species. Do, however, have a hygrometer, which measures humidity and a thermometer in view so you can check readily on temperature.

Misting systems to water plants are other equipment the avid gardener buys, but these are not really necessary either. Be sure there is a water outlet in the structure; water plants with a hose or a watering can. I find this much more satisfactory because then you have time to really see every plant as you water and to check for insects or fresh growth. With an automatic misting system there is no way to observe plants and to really enjoy each and every one of them.

I may seem to be against fancy equipment, and I am, but I do think that a small electric fan operating at low speed during hot weather is a must. These fans rarely cost more than a few dollars and are the best investment you can make to keep air moving. Few plants (people either) like a stagnant situation.

A small auxiliary heating unit sold at many department stores is a good idea to have on hand for those really cold nights. Sometimes even with solar helpers you will still need artificial heat, and the small heating units now on the market can be used (none is extremely expensive).

Lattice work is used as protection against noon summer sun in this area. It effectively breaks the sun and also looks well. (Photo by author).

Shading

Shading your greenhouse is important, so much so that it deserves a special section. In most summer climates you will need some form of shading to protect plants from intense noon heat that will burn them. If you have built a partial wood and glass or dome ceiling, little shading will be necessary, but if the ceiling consists of many glass or plastic domes, then shading is a must. In the old days shading was done with paste or powder with sprayer, paintbrush, or roller; it was always and still is a mess. The paste and powder has to be applied, often removed in winter, and always looks ugly.

It is a far better idea to trellis the greenhouse because this way the shading becomes a part of the greenhouse design and is esthetically pleasing. Trellis can be left in place all year and provide almost perfect light for most plants—alternate light and shade, a happy medium. A trellis can be easily made from redwood lath on frames, and it does look well.

Inexpensive wood or bamboo blinds (roll-up kinds) are another way to thwart the sun. These are made in very large sizes and sold at most Sears or large department stores. They are convenient and not as handsome as trellises, but they are satisfactory. Many people use plastic sheeting as protection from the sun. Mesh plastic or fiberglass is inexpensive and easy to install, but it is hardly pleasing to the eye. Still, plastic can certainly be used; put it up in the summer and remove it in the fall.

Another good protection from sun is wire screen. Make some wooden frames or buy some pre-made screens and put them in place. These are not unattractive and serve the purpose.

Actually, any minor problem can be corrected in the greenhouse once you determine just what that problem is—ventilation, humidity, temperature. First, know what is wrong. If it is too hot in summer in the growing area, more ventilation can be provided with windows or vents. If it is too cold in winter for plants, shutters and screens can be built to buffer winds and storms.

Ideally, none of these problems should occur if the room is properly built and with proper materials but no one is perfect and imperfections are bound to creep into the scheme of things. The point is—don't panic. There are always ways to remedy problems and make the greenhouse a thriving, suitable place for plants.

Bibliography

Architecture Through the Ages, Talbot Hamlin, G. P. Putnam's Sons, 1940

Art of Home Landscaping, Garett Eckbo, McGraw Hill, 1956

Climate and Architecture, Jeffrey Aronin, Reinhold Publishing, 1953

Climate and Man, Yearbook of Agriculture, 1941, United States Government Printing Office, Washington, D.C.

Cooling Effects of Trees and Shrubs, University of California (bulletin), Davis, Ca., 1958

Dome Builders' Handbook, John Prenis, Running Press, Philadelphia, Pa., 1973

Garden Ornament, Gertrude Jekyll, Country Life Publishers, 1938

In the Nature of Materials/Buildings of Frank Lloyd Wright, Henry Russell Hitchcock, Duell, Sloan, and Pearce, 1942

New Roofs for Hot Dry Regions, Harold R. Hay (Reprinted from EKISTICS), Volume 3, Number 183, Feb. 1971

Owner Built House, Ken Kern, Homestead Press, 1972

Shelter, Shelter Publications, Bolinas, Ca., 1973

Solar Era- Mechanical Engineering Magazine, Harold R. Hay, Oct. 1972

Solar Energy, Solar Power and Pollutions (Harold R. Hay), International Congress, Paris, 1973

Winter Flowers in Greenhouse and Sun Heated Pit, Kathryn S. Taylor and Edith W. Gregg, Chas. Scribner's Sons, 1969

Zome Primer, Steve Baer, Zomeworks Corp., 1970